Arabian Wolf

# The Mammals
## of Israel

**Other titles by the author**

*Living Animals of the Bible* (illustrated by the author)

*Birds of Our Land* (illustrated by the author)

*Mammals of Our Land* (illustrated by the author)

*On the Snows of Mount Hermon* (co-author and co-illustrator)

# The Mammals of Israel

Written and Illustrated by

## Walter W. Ferguson

Edited by

## Susan Menache

gefen
publishing house בית הוצאה לאור
JERUSALEM ◆ NEW YORK

Typesetting: Marzel A.S. – Jerusalem
Cover production: Studio Paz, Jerusalem
Cover design, Illustrations: Walter W. Ferguson
Edited by Susan Menache

1  3  5  7  9  8  6  4  2

Gefen Publishing House
POB 36004, Jerusalem 91360, Israel
972-2-538-0247 • orders@gefenpublishing.com

Gefen Books
12 New Street Hewlett, NY 11557, USA
516-295-2805 • gefenbooks@compuserve.com

**www.israelbooks.com**

Printed in Israel                                    *Send for our free catalogue*

---

ISBN 965-229-278-8

**Library of Congress Cataloging-in-Publication Data:**
Ferguson, Walter
The Mammals of Israel: Written and Illustrated by Walter W. Ferguson; Edited by Susan Menache.
Includes bibliographical references (p. )
1. Mammals—Israel.  I. Menache, Susan.  II. Title.
QL729.I75 F48 2002 • 599'.095694—dc21 • CIP Number: 2001051103

# CONTENTS

# COLOR PLATES

# BLACK AND WHITE FIGURES

# THE MAMMALS
# OF ISRAEL

# FOREWORD

Walter Ferguson and I grew up in Brooklyn within walking distance of each other. We were almost the same age but went to different high schools — he to Midwood and I to James Madison — and might never have met had it not been for a teacher at each of the schools. At Midwood, Abe Rabinowitz taught Biology; at Madison, his wife, Anne, taught English and Speech. With Abe's enthusiastic encouragement, Walter painted a large, accomplished mural depicting many local birds on a wall at Midwood. I had produced numerous pencil drawings of saturniid moths that, through Anne's connections, were matted and exhibited along several of Madison's hallways. Walter's mural and my exhibition were exactly contemporaneous and both caused something of a sensation at the two schools. Anne and Abe, of course, arranged for Walter and me to meet. Abe took me to see Walter's mural. I was bowled over. That a boy 16 years old could handle paint with such precision and bravura, and had such drawing skill and mastery of aerial perspective, seemed almost supernatural to me. I still don't know how he did it. With much in common, we became life-long friends. The two wonderful teachers remained like family to both of us until their much lamented deaths many years later.

A few years later, I had the opportunity to share a big project with Walter and work side by side with him in his tiny rented studio. He had been engaged by the C.L. Hammond map company to produce over 300 small paintings for their *Nature Atlas of America* — a monumental task. As his deadline approached, he saw that he might need help to finish. He contacted me, and the result was that I turned out 38 paintings of insects.

It was at that time that I became fully aware of Walter's great breadth of interests and the unusual range of his skills. His particular genius was not like Audubon's, limited to the painting of birds and animals (Audubon did paint other things, though not very well.) Even Walter's backgrounds, whether landscapes or seascapes, were accomplished with similar mastery.

Walter Ferguson, as a young man, was one of only a handful of artists who painted the natural world in a representational, respectful and admiring way. He truly loved the things he painted. His teenage painting of the bird mural was done at the very height of the exaltation of abstract expressionism. The art establishment derided as kitsch all realistic painting, and anyone who painted recognizable subjects with sensitivity, discrimination and subtlety — as Ferguson did — was denigrated as lacking in artistic sense and devoid of creative talent. No one knows how many able young artists threw in the towel for lack of encouragement and either ceased to paint or began imitating Picasso, but the number must be vast. It is hard to persist in a vocation in the face of what seems the depreciation of the whole civilized world. Ferguson, to his great credit, refused to capitulate and never abandoned his inner-directed standards, continuing over the decades to turn out works of genuine beauty and truth. So-called wildlife art, of which Ferguson was a forerunner, is now widely popular, though still the object of mockery by the critics and curators who, themselves, are rarely artists and most often the products, not of ateliers, but of academia. How the mantle of absolute aesthetic authority was bestowed on them is something of a mystery. Realistic art, they proclaim, is "mere illustration". "Illustration" and "illustrator" have now become almost pejorative terms. What is ignored is that almost all Western art and much Eastern was, until the 20th century, exactly that — illustration, the Sistine ceiling being perhaps the most effulgent example among myriad masterpieces.

When unmade beds and the corpses of calves in formalin solution are considered art — as they are now by influential critics and curators — it is surely long past time for a revision of standards.

As an artist Ferguson is not easily pigeon-holed. His versatility and knowledgeable grasp of so many kinds of subject matter may, in fact, have worked against him when it came to recognition. Most artists stake out circumscribed areas of art and stick to it, becoming known as bird artists, landscape painters, genre painters, etc. Art critics like that; it makes their life and work easier for them. It is paradoxical that Ferguson's very strengths may, it seems, be a major cause of his escaping the critical notice he deserves.

Walter Ferguson is one of those artists who has paved the way. His earlier books, with their lovely and accurate paintings of birds and mammals, have been an inspiration to aspiring young artists, providing the example and encouragement so completely lacking in his own youth.

Human courage takes many forms. If persisting in the face of almost insurmountable obstacles is a form of courage, as most of us think it is, then Walter Ferguson is a man and artist to be greatly admired and appreciated.

John Cody
June 5, 2001

# PREFACE

The Bible is our first written record of the mammals that lived in Israel during historic times. Although the philology is not always clear, more than 20 species have been identified.

When I first visited Israel in 1958, there were two books dealing with the mammals of Israel: one was *The Fauna and Flora of Palestine*, by H.B. Tristram, 1884, and the more recent *Animal Life in Palestine*, by F.S. Bodenheimer, 1935. Although most of the species in the country were already known, there was little information available about subspecies, their habits and distribution. The book was poorly illustrated in black and white and not very informative. *The Mammals of Arabia* by David Harrison, published in 1964, 1968, 1972 and 1991, contributed greatly to our knowledge, correcting some previously-held misidentifications, as well as describing new species and subspecies. However, it was primarily a taxonomic work, and contained only black and white illustrations. There were still taxonomic problems to be resolved, and much to be learned about habits, habitat and distribution. Moreover, some species had already become extinct under Turkish and British rule; others were endangered. In short, I felt that there was a need for a book on the mammals of Israel.

In 1966, a year after immigrating to Israel, I was appointed staff artist of the Department of Zoology of Tel Aviv University, where, in the course of my work, I learned the art of taxonomy. The university had a research zoo and a museum with a collection of native mammals, to which I added specimens from time to time. I was privileged to have access to these facilities, and so it was possible for me to become intimately acquainted with the mammals of Israel. During this period, I wrote and illustrated a book entitled *The Living Animals of the Bible* and the first guide to the mammals of Israel, *Mammals of Our Land*, which although not comprehensive, created an increased awareness of the mammals of Israel. I also co-authored and co-illustrated a book on the flora and fauna of Mount Hermon called *On the Snows of Mount Hermon*.

Over the course of the last 44 years, I have collected and studied the mammals of Israel. This was made possible by assistance I received from the Tel Aviv University, the Nature Reserves Authority, the Nature Protection Society of Israel, the Hebrew University of Jerusalem, Beit Ussishkin Museum and other insitutions, as well as many individuals. The goal in mind was to write and illustrate in color a book of all the species and subspecies of the mammals of Israel.

In 1993, a Field Guide to the Mammals of

Israel by B. Shalmon was published in Hebrew. Unfortunately, it is not comprehensive and very poorly illustrated. Other pertinent publications are a volume on mammals for the *Nature Encyclopedia of Israel* and *Mammalia of Israel* by H. Mendelssohn and Y. Yom-Tov (1999), illustrated with color photographs, and a book entitled *Mammals of the Holy Land* by M.B. Qumsiyeh, published in 1996, which contains only black and white photographs of many, but not all species.

In most books published to date, including those mentioned above, the illustrations are secondary to the text. Photographs often do not show the animal in a characteristic position or its correct color. In this book, the major contribution is more than 165 illustrations. These are not meant to be simple, diagramatic illustrations for identification, but portraits that capture the character, as well as the color, of each species. Some subspecies, are shown for the first time.

The text is mainly descriptive of those external features useful in identification, habits, habitat and distribution in general, and of the particular races found in Israel. It contains some taxonomic remarks regarding the systematic position of certain species and races. For a detailed description of external, cranial and dental features, I refer the reader to *The Mammals of Arabia* by David Harrison (1991) and for behavior, *Fauna Palaestina Mammalia of Israel* by Heinrich Mendelssohn and Yoram Yom-Tov (2000).

Referring to the importance of illustration, Mayr et al. (1953) stated that "words, no matter how carefully chosen, are rarely adequate to give an accurate mental picture of the appearance of an organism." There is also an old saying: "A picture is worth a thousand words." Alan Walker, a scientist, and his headmaster claim that "you can always be a scientist and do art as a hobby, but you can never be an artist and do science as a hobby." (Walker & Shipman, 1996). What then was Leonardo de Vinci? Referring to Leonardo, MacCurdy (1939) states that "this habit of scientific investigation in inception subsidiary to the practice of his art, grew so as to dominate it..."

I am very grateful to my late wife, Grace, who loved animals and tolerated the menagerie of small mammals I trapped for study and painting; for her patience while I stopped and photographed road-kills; and for allowing me to keep specimens in the freezer. A special thanks goes to my wife, Ruth, who did not realize what she was getting into by marrying a wildlife artist, and without whose generosity this book would not have been possible.

This book was a labor of love and I want to thank many people, friends and colleagues for their assistance: Mr. Mike Cohen, Prof. Heinrich Mendelssohn, Prof. Zvi Yaron, Prof. Yehoshua Kugler, Mr. Uri Marda, Ms. Tsila Shariv, Mr. Giora Ilani, Prof. Ilan Golani, the late Mr. Elimelech Hurvitz, Dr. David Makin, Prof. Yehuda Wharman, Dr. Shlomo Hellwing, Dr. David Harrison, Mr. Michael Greivenbrock, Ms. Hadass Laferriere, Ms. Susan Menache and Robert Hoffmann.

Walter W. Ferguson

# INTRODUCTION

When man was a hunter/gatherer, he lived in harmony with nature. But since the beginning of agriculture, man and wild animals have been incompatible. Large predators were the first to go. Many lions were exported to Rome to fight gladiators. Apart from a few commensal species, such as mice, rats and bats, many mammals have been competing with man for the use of natural resources. As human numbers have increased and spread, the mammal population has decreased and retreated. Man was once surrounded by wild animals. Today, wild animals are surrounded by man.

It is therefore surprising that mammals, such as the leopard and wolf, have survived in Israel until the present day, when they have become extinct in much larger countries. This is probably due to a combination of several factors.

Firstly, although the land of Israel has been inhabited and exploited by man for hundreds of thousands of years, it has lain relatively fallow for centuries with a sparse human population. The Negev Desert and Dead Sea depression, which are home to the leopard and wolf, were, until comparatively recently, largely innaccessable, except to some Beduin tribes. It is only in the past several decades that man has made inroads, literally, into the territory of the leopard and the wolf and consequently threatened their survival.

In Israel, archaeological remains testify to the transformation from hunting and gathering to agriculture and the domestication of animals. Gazelles were hunted on the plains and flat valleys, Aurochs and deer in the forests, and ibex on the mountain ridges. About 12,000 years ago, the Aurochs, (*Bos taurus*), Bezoar Goat (*Capra aegagrus*) and wolf were first domesticated. The wolf, not the jackal, is ancestral to the dog. There is osteological evidence that the Fallow Deer, (*Dama dama*), lived in Israel during the Bronze Age (Ferguson et al., 1985).

Secondly, advanced technology is not without its cost. Old hunting methods by rifle on horseback or camel did not endanger the existence of a species, but modern automatic weapons used from vehicles with a four-wheel drive have virtually wiped them out. Fifty years ago, the Sinai Ibex and two species of gazelles were endangered species in Israel. They had been hunted to the verge of extinction, and were only saved by strict government protection.

Moreover, the traffic situation has changed drastically. Forty years ago, the roads in Israel were narrow, and the cars few in number. The narrow roads constituted only a small interruption in the environment, and crossing the gap resulted in many road-kills — animals hit by cars. Today, the roads are wide — up to six lanes of speeding cars. Few mammals attempt to cross such a barrier, and so there are far fewer road-kills. However, criss-crossing superhighways and other barriers have fragmented the ecosystem. In the long run, this causes populations to become isolated from each other, which weakens them genetically.

A further great blow to the mammalian fauna was the extensive destruction of natural habitat. Beginning in biblical times, agriculture gave way to pastoral overgrazing by goats, sheep and camels. The felling of trees for construction by the Romans and charcoal and lime burning have destroyed most of the forests and shrub lands in Israel, resulting in widespread erosion. In addition, frequent fires were due to irresponsible shepherds. Under the Ottoman Empire, the need to fuel the railroad led to widespread deforestation. These were cut down to avoid taxes. The Lynx, (*L. lynx*), may have disappeared with the forests. The little Roe Deer and Mesopotamian Fallow Deer gradually died out in the northern half of the country where they had lived for hundreds of thousands of years. Today, only isolated Allepo Pines survive as relics of former pine forests succeeded by oaks. On Mount Carmel, the Beech Marten disappeared and retreated northward on the heels of the squirrel, its main prey. Citrus and olive groves replaced the oak forests, and now urban sprawl is replacing the citrus groves. The desert, meanwhile, is insiduously creeping northward.

The destruction of natural habitat is the most important factor threatening the survival of many species. In a small, developing country, such as Israel, agricultural land is needed. And so the great Huleh Lake and swamp was drained, as well as other swamps in the Jezreel valley and on the coast. In those days, little was known about the consequences of interfering with the balance of nature. There was no such thing as an environmental impact study. As a result, the swamps were drained, to the detriment of the wildlife that lived there. Several species of plants and animals became extinct, including the Syrian Water Vole, and the Common Otter has became scarce due to diversion and pollution of the rivers.

The desert, thought by many as a wasteland, is not all sand, but a delicate and fragile ecosystem that has taken thousands of years to reach its present climax of specially-adapted flora and fauna. Israel has "made the desert bloom" by replacing the native Tamarisks with Date Palms and fruit trees, but when the desert is disturbed, the natual flora cannot be restored. Even after 50 years or more, a great scar will remain on the land. The Acacias are gradually disappearing, and so will the gazelles who feed on them.

Israel is the only country in the region where large predators, such as the wolf, leopard and hyaena, are protected. However, in spite of this, a battle between the farmer and predators is being waged. One farmer, who took matters into his own hands after some cattle were killed by wolves, laced a calf and chickens with poison as bait. The known death toll among wild animals reached 27 Griffin Vultures, three Egyptian Vultures, seven Syrian Jackals, three Palestine Wild Boars, one Palestine Wild Cat and one Domestic Cat. Not a single wolf was killed. Pesticides also kill rodents which are then eaten by larger animals, who die from secondary poisoning. Jackals are poisoned because of the fear of rabies and have been exterminated from parts of the country.

The Nature Reserves Authority and the Israel Society for the Protection of Nature are doing their best to protect wildlife. The return of the Sinai Ibex and Mountain Gazelle in numbers shows that a reversal is possible. However, land is valuable, and economics is the bottom line. Conservation does not come free; it takes money to preserve natural habitats, and it is the developers and not "tree huggers" who have the money. Education may be the answer; but unfortunately, people are increasing in numbers faster than they are being educated.

Introducing invasive species, such as rats, dogs, cats and goats, into new areas causes disorder in the ecosystem by spreading disease and killing native species.

In conclusion, due to the population explosion and its consequences, more people means the development of more land, which means increased environmental impoverishment. The outlook for the future of mammals in Israel is bleak. This does not just apply to the predators; small creatures, such as gerbils and jerboas, are also in danger as long as coastal sand dunes are turned into garbage dumps or parking lots. Animal reserves and safari parks are all well and good, but they are really nothing more than glorified zoos. At this rate, the day will soon come when a Beduin child has to go to a zoo to see a camel.

Although we cannot easily see mammals in the wild, except in Africa, we know that they are there. We hear them at night, and in the morning we see their footprints or find them dead on the road, having been hit by a car. There are some that are active during the day, such as the Syrian Rock Hyrax, Egyptian Mongoose and Sinai Ibex. But for the most part, daytime belongs to man, and nighttime to mammals. It is their nocturnal nature which has helped them to survive. They live in harmony with the world, while we manipulate and dominate them.

Today, more species of animals are becoming extinct and at a faster rate than ever before in history. In the past few thousand years, no less than ten species of mammals alone have vanished from the land of Israel. That is more than 10 per cent of the total number that remain. We have entered a new age of impoverished biological diversity and mass extinction — by mankind.

One of the first inspirations for art was mammals from the famous cave drawings in France and Spain by early man. Ancient rock drawings and engravings of animals are widespread from the Sahara Desert to South Africa and Australia. More recent etchings on rocks in the desert of Israel depict camels, oryx, ibex and horses. During the long hunter/gatherer period of mankind's existence, he had close contact with wild animals. Throughout the development of agriculture and rural life, man still lived close to nature and was familiar with the animals around him. Images of animals became important to his culture. The lion was the symbol of the tribe of Judah.

With the advent of the industrialized world, however, people moved into cities and their personal contact with animals virtually disappeared, along with familiarity with them. There are zoos and safari parks, but it was television and books that brought animals back into the life of the man in the street. These images remind us of the amazing diversity of animals with which we share the earth and the wonder, knowledge and enjoyment they give us.

Bodenheimer (1960) pointed out the taboo regarding the representation of many animals throughout the Middle East. Big game animals, those symbolic of religion, of economic interest and exotic species first roused the interest of artists. Yet the provincial character of the civilization in the region barred artistic creativity. Another factor is the religious admonition by Jews and Arabs against creating graven images of man or animal, and the refractory attitude towards most of the fine arts. To this day, Israel is far behind the rest of the world in the appreciation of paintings of animals as fine art.

It is to be hoped that this book will increase an awareness of mammals, enhance appreciation of Israel's wonderful wildlife heritage, and the importance of protecting it for future generations. I believe that the survival of mammals is in our hands. I hope that we are ready to take on that responsibility — before it is too late. We will never regain what is lost.

# IDENTIFICATION

Most of the species of mammals in Israel are easily identified by their external characters. However, some of the small species may require cranial and dental measurements to make a definitive identification. Certain questions should be asked when identifying a specimen. Is a difference in size or robusticity due to age, sex and normal variation since no two specimens are exactly alike? Older specimens and males are usually more robust. Populations living in colder cimates tend to be larger and more robust than those in warmer climates and with shorter ears. As for coloring, there is often great variation due to age, the young being different from the adults; a seasonal difference in pelage, shorter in the summer and longer in the winter; the reflection of substrate coloration, darker in humid areas, paler in arid areas; and due to wear or fading. In some species there are different color phases, such as black or red. Light-colored mutants occur in dark-colored populations and vice versa. Finally, a specimen may be a hybrid, a cross between two subspecies, and, in such a case, it usually does not fall completely into either subspecies.

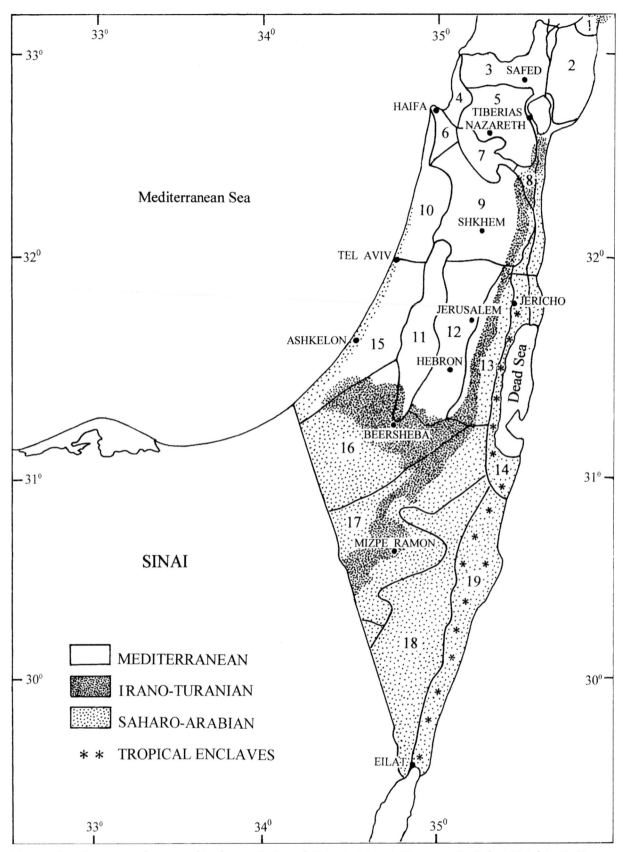

Zoographical zones and geographical areas in Israel. Key: 1. Mount Hermon; 2. Golan Heights; 3. Upper Galilee; Northern Coastal Plain; 5. Lower Galilee; 6. Carmel Ridge; 7. Yizr'el Valley; 8. Southern Golan Heights and Jordan Valley; 9. Samaria; 10. Central Coastal Plain; 11. Foothills of Judea; 12. Judean Hills; 13. Judean Desert; 14. Dead Sea area; 15. Southern Coastal Plain; 16. Northern Negev; 17. Central Negev; 18. Southern Negev; 19. Arava Valley.

# ZOOGEOGRAPHY

Israel is a small strip of land wedged between Sinai and Arabia. It is 20,600 km² — about the size of New Jersey. To the north, it borders on the mountains of Lebanon; to the east, the Golan Heights, the Jordan River, the Dead Sea and the mountains of Edom; to the west, the Mediterranean Sea and the Sinai Desert; and at its southernmost tip, the Gulf of Akaba, an arm of the Red Sea. The country is remarkably varied topographically in a relatively small area: mountains 2,000 m. above sea level in the north; bounded by the rift valley to the east; with the Dead Sea 400 m. below sea level; a broad maritime plain to the west; and the Negev Desert in the south.

Israel has been a crossroads for both animals and people from three continents: Europe, Asia and Africa. There are four zoographical zones represented in Israel: the Mediterranean Region, the Irano-Turanian Region, the Saharo-Arabian and tropical enclaves (Fig. 1). Consequently, Israel possesses a unique combination and diversity of mammals, unlike anywhere else in the world. The large number of species may be attributed to its position as a transitional zone.

However, it was not always so. To understand and appreciate the present day mammalian fauna of Israel and its distribution, we must delve into the geological and climatic history that molded the region.

Let us begin with the Cretaceous period, when Israel was covered by the sea. By the time of the Eocene period, the ancient Mediterranean Sea, called the Tethys Sea, retreated from Israel, leaving the land linked with Africa and Arabia, but separated from Eurasia. The primitive hedgehog already existed, and the ancestor of the Red Sea Dugong (*D. dugong*) was common from Egypt to northern Germany.

Next came the Oligocene epoch, which was hot and humid. The mammals of Israel in that period were similar to those of the African tropics.

The Miocene epoch saw the uplifting of mountains and the formation of the Great Rift, a chain of valleys extending all the way from what is today Syria, through the Jordan Valley, the Dead Sea region and the Arava Valley to East Africa. Israel was still broadly connected to Africa by way of Sinai, and with Arabia, but separated from Eurasia by the Tethys Sea. The climate was tropical, with a prehistoric fauna of grazing and cursorial animals inhabiting a Sudannean type of savannah in the south. Among these were the ancestors of today's antelopes.

In the late Miocene, the Tethys Sea became almost completely dry except for some brackish lakes. A land bridge now connected Africa with Eurasia. Although this allowed for an interchange of fauna, most of the movement seemed to be in the direction from Asia to Africa. The Trident Leaf-nosed Bat, Crested Porcupine and Short-tailed Bandicoot Rat spread from the tropical oriental region to the tropical Ethiopian region, while the range of the lion, leopard, cheetah and Mountain Gazelle extended from Africa to Asia.

During the late Miocene and early Pliocene epochs, there was a gradual drying up of the subtropics, indicating the beginning of the Ice Age. A great desert appeared in North Africa and extended across Sinai, through Arabia, southern Iran, west Pakistan and northwest India. The emerging desert broke the connection between the tropical fauna of Asia and Africa, and animals living there had to migrate, adapt, or perish.

The desertification had a deleterious effect on the mammals. Some species of African origin disappeared, namely the Ethiopian Giraffe, (*Giraffa camelopardalis*); Bubal Hartebeest, (*Alcelaphus buselaphus*); and the Crested Rat, (*Lophiomys imhausi*). The elephant was able to migrate, but the hippopotamus, (*Hippopotamus amphibius*), perished. The last Bubal Hartebeest was shot around 1876. With the desertification of the Negev and the Arava Valley, the leopard, jackal, Sinai Rock Hyrax and Short-tailed Bandicoot Rat withdrew to tropical enclaves or oases where they became relics to this day.

Although Tristram reported the Addax (*Addax nasomaculatus*) from the desert of southern Palestine, it has never been recorded east of the Nile. Similarly, there are claims for the existence of the Arabian Oryx (*Oryx leucoryx*) west of the mountains of Edom in Jordan, but this species is adapted to a sandy desert and not likely to have crossed the mountainous barrier into western Palestine. Its occurrence in Israel has not been substantiated with hard evidence. It has nevertheless been introduced into the Arava Valley of Israel by the Nature Reserves Authority.

Mammals did not enter the desert. They were already present there when the region was a savannah or steppe and evolved into desert forms by adapting to the change in climate and environment. For example, the Lesser Three-toed Jerboa derived from the Four- and Five-toed Jerboa. Many grew smaller and paler, with inflated tympanic bullae. The Cape Hare became smaller with relatively larger ears. Some species, such as the Sand Cat and Rueppel's Sand Fox, grew hairy soles for better traction on the sand. The desert produced a whole new fauna, mostly of African origin, highly adapted to the hostile arid conditions. The true desert species, such as the Dorcas Gazelle, Golden Spiny Mouse and jerboas get enough water from the food they eat and can survive without free drinking water. The climate during the Pliocene was not unlike that of today, and most of the species now found in Israel probably had their origin in the Pliocene.

A number of mammals in Israel have been regarded "indigenous species that evolved in isolation" (Harrison, 1964). These are considered here to be no more than local subspecies. They are the Arabian Barbastelle, Bodenheimer's Pipistrelle, Arabian Hare, Levant Garden Dormouse, Allenby's Gerbil, Cheesman's Gerbil and Buxton's Jird.

In the late Pliocene and early Pleistocene epochs, the Jordan-Orontes Rift Valley, with its chain of lakes and swamps, apparently became a major flyway for migrating birds. Many of these are aquatic. Even though the lakes and swamps have long since gone, they still continue to take this ancient flyway across the desert. Fossil remains of elephants from this period were found in Bethlehem.

The Pleistocene Ice Ages had their influence on the mammals of Israel even though the ice cap never reached that far south. Among the early Pleistocene fauna was the Barbary Ape, (*Macaca sylvana*), which still survives in North Africa. The Brown Hyaena (*H. brunnea*) and the *Stegadon*, intermediate between the elephant, mastadon and rhinoceros (*Dicerorhinus*), lived until the mid-Pleistocene. During the course of three or four successive glaciations, Eurasian fauna was pushed south by the advancing cold and snow. During these periods, the climate grew cooler and wetter. Remnants of a juniper forest in northern Sinai desert give mute testimony to these climatic changes.

The hyrax, which was widespread in the Pliocene from southern Europe to China, retreated south before the advancing Ice Age to the Middle East and Africa. During this period, some Eurasian fauna entered Arabia as far south as Yemen. The Arabian Barbastelle reached its southern limit in Israel near Eilat.

In the inter-glacial periods, many species returned north in the wake of the retreating ice, while others, such as the Alpine Field Mouse and the Snow Vole, remained and survived by withdrawing to higher, cooler elevations. Some allopatric populations became new races (subspecies). Where the trees disappeared, the normally arboreal Southwest Asian Garden Dormouse became terrestrial. Eurasian species, such as the wolf, jackal, ibex and Mountain Gazelle, which had spread as far south as Saudi Arabia, adapted to the desert. The ubiquitous Cape Hare evolved in a cline running north/south from Eurasia to South Africa.

The climate of Israel changed several times during the Pleistocene, with alternate wet and dry periods. During the period from 8000 to 1000 BCE, the Aurochs or Wild Ox, inhabited Israel. About 7,000 years ago, the Bezoar or Wild Goat, ancestor of the domestic goat, disappeared from the north, while the Sinai Ibex survived in the south. The Red Deer (*Cervus elaphus*) survived until 3,000 years ago. Although the Barbary Sheep, (*Ammotragus lervia*), a Saharan species, is mentioned in the Talmud, and reported to have been seen by Aharoni, south of the Dead Sea (circa 1903), there is no material evidence of its presence in Israel. An isolated colony of the Indian Elephant, (*Elaphus maximus*), survived in Syria until about 800 BCE. The fauna of the maritime plain of Israel was contiguous with that of Egypt and North Africa, but during the last 6–10,000 years, there has been a gradual drying up. The coastal plain of northern Sinai was inundated by wind-blown sands of the Libyan and eastern deserts. As a result, the northern coastal plain of Sinai has become the only place in the circum-Mediterranean region where the desert meets the sea, and is known as the Sinaitic-Libyan gap.

The Sinaitic-Libyan gap is a subregion of the Saharo-Sindian region. It consists of great shifting sand dunes, with scattered vegetation, oases and sterile salt flats. The intrusion of the Sinaitic-Libyan gap has destroyed the Mediterranean fauna that existed in the coastal plain of northern Sinai. The discontinuous geographic distribution of numerous species is evidence of the succession of a whole fauna in recent times. Many water-dependent species found in the coastal plain of Israel, Egypt and western Libya are significantly absent from the northern coastal plain of the Sinai Desert. Among these are the Egyptian Mongoose, Wild Boar, Asiatic Jackal, European Hedgehog, Greater Mole-rat and Social Vole. Some species, such as the Great and Lesser White-toothed Shrews and Kuhl's Pipistrelle, have survived in microclimates of oases in this desert-like habitat. Others have evolved into new subspecies, such as Buxton's Jird, Schlueter's Lesser Three-toed Jerboa and Allenby's Gerbil.

The expansion of the Sinaitic-Libyan gap across northern Sinai and up the southern coast of Israel has brought in its wake a Saharan flora and fauna. Migrating islands of sand permitted psammophiles, such as the Great and Lesser Egyptian Gerbil, to penetrate the northwestern Negev Desert.

As the sand dunes accumulated along the coast, gerbils, jirds and jerboas spread north. During times of drought, when the mouths of coastal rivers became blocked, these rodents were able to cross the dry river beds, but when the rivers swelled from heavy rains and the rushing waters broke through to the sea during prolonged wet periods, these populations became isolated. As a result, endemic races appeared, such as Schleuter's Lesser Three-toed Jerboa and Allenby's Gerbil, the former reaching close to Tel Aviv, and the latter as far north as Haifa. Two races of the Dorcas Gazelle entered Sinai from Egypt across the Isthmus of Suez long before the canal was built. One came from the south around the Red Sea littoral, while the other came from the desert in the north. The two subspecies, which had been separate in Egypt, met and interbred in Sinai, and a salient hybrid population spread into southern Israel.

As a final word, the climate appears once again to be warming up and if prolonged, we may expect it to have its effect on the present fauna of Israel, as it has in the past.

# SYSTEMATIC LIST

96 extant endemic species

6 extirpated in historic times

3 introduced feral species

2 reintroduced species

| | | |
|---|---|---|
| **Order** | **INSECTIVORA: INSECTIVORES** | |
| Family | ERINACEIDAE: HEDGEHOGS | |
| Genus | *Erinaceus* | |
| | *Erinaceus europaeus concolor* | Eastern European Hedgehog |
| Genus | *Hemiechinus* | |
| | *Hemiechinus auritus aegyptius* | Egyptian Long-eared Hedgehog |
| Genus | *Paraechinus* | |
| | *Paraechinus aethiopicus pectoralis* | Ethiopian Hedgehog |
| | | |
| Family | SORICIDAE: SHREWS | |
| Genus | *Suncus* | |
| | *Suncus etruscus etruscus* | Pygmy Shrew |
| Genus | *Crocidura* | |
| | *Crocidura leucodon judaica* | Bicolor White-toothed Shrew |
| | *Crocidura lasia* | Thomas's White-toothed Shrew |
| | *Crocidura russula monacha* | Great White-toothed Shrew |
| | *Crocidura suaveolens portali* | Lesser White-toothed Shrew |
| | *Crocidura ramona* | Ramon White-toothed Shrew |
| | *Crocidura pergrisea* | Pale Gray Shrew |

| | | |
|---|---|---|
| **Order** | | **CHIROPTERA: BATS** |
| Suborder | | MEGACHIROPTERA: FRUGIVORUS BATS |
| Family | | PTEROPODIDAE: FRUIT BATS |
| Genus | *Rousettus* | |
| | *Rousettus aegyptiacus aegyptiacus* | Egyptian Fruit Bat |
| Suborder | | MICHROCHIROPTERA: INSECTIVORE BATS |
| Family | | RHINOPOMATIDAE: MOUSE-TAILED BATS |
| Genus | *Rhinopoma* | |
| | *Rhinopoma microphyllum microphyllum* | Larger Mouse-tailed Bat |
| | *Rhinopoma hardwickei cystops* | Lesser Mouse-tailed Bat |
| Family | | EMBALLONURIDAE: SHEATH-TAILED BATS |
| Genus | *Taphozous* | |
| | *Taphozous nudiventris nudiventris* | Naked-bellied Tomb Bat |
| | *Taphozous perforatus haedinus* | Tomb Bat |
| Family | | NYCTERIDAE: SLIT-FACED BATS |
| Genus | *Nycteris* | |
| | *Nycteris thebaica thebaica* | Egyptian Slit-faced Bat |
| Family | | RHINOLOPHIDAE: HORSESHOE BATS |
| Genus | *Rhinolophus* | |
| | *Rhinolophus f. ferrum-equinum* | Greater Horseshoe Bat |
| | *Rhinolophus clivosus clivosus* | Desert Horseshoe Bat |
| | *Rhinolophus hipposideros minimus* | Lesser Horseshoe Bat |
| | *Rhinolophus blasii* | Peter's Horseshoe Bat |
| | *Rhinolophus euryale judaicus* | East Mediterranean Horseshoe Bat |
| | *Rhinolophus mehelyi* | Mehely's Horseshoe Bat |
| Family | | HIPPOSIDERIDAE: LEAF-NOSED BATS |
| Genus | *Asellia* | |
| | *Asellia tridens tridens* | Trident Leaf-nosed Bat |

| | | |
|---|---|---|
| Family | | MOLOSSIDAE: FREE-TAILED BATS |
| Genus | *Tadarida* | |
| | *Tadarida teniotis rueppelli* | European Free-tailed Bat |

| | | |
|---|---|---|
| Family | | VESPERTILIONIDAE: PLAIN-NOSED BATS |
| Genus | *Nyctalus* | |
| | *Nyctalus noctula lebanoticus* | Common Noctule |
| Genus | *Myotis* | |
| | *Myotis myotis macrocephalicus* | Greater Mouse-eared Bat |
| | *Myotis blythi omari* | Lesser Mouse-eared Bat |
| | *Myotis emarginatus emarginatus* | Geoffroy's Bat |
| | *Myotis capaccinii bureschi* | Long-fingered Bat |
| | *Myotis nattereri hoveli* | Natterer's Bat |
| Genus | *Eptesicus* | |
| | *Eptesicus serotinus serotinus* | Serotine Bat |
| | *Eptesicus bottae innesi* | Botta's Serotine Bat |
| Genus | *Pipistrellus* | |
| | *Pipistrellus pipistrellus pipistrellus* | Common Pipistrelle |
| | *Pipistrellus kuhli ihkwanius* | Kuhl's Pipistrelle |
| | *Pipistrellus rueppellii coxi* | Rueppel's Pipistrelle |
| | *Pipistrellus savii caucasicus* | Savii's Pipistrelle |
| | *Pipistrellus savii bodenheimeri* n. status | Bodenheimer's Pipistrelle |
| | *Pipistrellus ariel* | Desert Pipistrelle |
| Genus | *Otonycteris* | |
| | *Otonycteris hemprichii jin* | Hemprich's Long-eared Bat |
| Genus | *Barbastella* | |
| | *Barbastella barbastella leucomelas* n. status | Arabian Barbastelle |
| Genus | *Plecotus* | |
| | *Plecotus austriacus christiei* | Gray Long-eared Bat |
| Genus | *Miniopterus* | |
| | *Miniopterus s. schreibersi* | Schreiber's Bat |

## Order CARNIVORA: CARNIVORES

| | | |
|---|---|---|
| Family | | CANIDAE: DOGS, WOLVES, JACKALS AND FOXES |
| Genus | *Canis* | |
| | *Canis lupus pallipes* | Indian Wolf |
| | *Canis lupus arabs* | Arabian Wolf |
| | *Canis aureus syriacus* | Syrian Jackal |
| | *Canis aureus hadramauticus* n. status | Arabian Jackal |

| | | |
|---|---|---|
| Genus | *Vulpes* | |
| | *Vulpes vulpes* | Common Red Fox |
| | *Vulpes vulpes flavescens* | Mountain Red Fox |
| | *Vulpes cana* | Blanford's Fox |
| | *Vulpes vulpes niloticus* | Egyptian Red Fox |
| | *Vulpes vulpes arabica* | Arabian Red Fox |
| | *Vulpes vulpes palaestina* | Palestine Red Fox |
| | *Vulpes rueppellii rueppellii* | Rueppel's Sand Fox |
| | *Vulpes rueppellii sabaea* | Rueppel's Sand Fox |

Family            URSIDAE :BEARS

| | | |
|---|---|---|
| Genus | *Ursus* | |
| | *Ursus arctos syriacus* | Syrian Brown Bear |

Family            MUSTELIDAE: WEASELS, MARTENS,
                                 POLECATS, BADGERS AND OTTERS

| | | |
|---|---|---|
| Genus | *Mustela* | |
| | *Mustela nivalis* | Snow Weasel |
| Genus | *Martes* | |
| | *Martes foina syriaca* | Syrian Beech Marten |
| Genus | *Vormela* | |
| | *Vormela peregusna syriaca* | Syrian Marbled Polecat |
| Genus | *Meles* | |
| | *Meles meles canescens* | Persian Badger |
| Genus | *Mellivora* | |
| | *Mellivora capensis wilsoni* | Honey Badger |

Subfamily          LUTRANAE: OTTERS

| | | |
|---|---|---|
| Genus | *Lutra* | |
| | *Lutra lutra seistanica* | Common Otter |

Family            VIVERIDAE: MONGOOSES, CIVETS AND GENETS

Subfamily          HERPESTINAE: MONGOOSES

| | | |
|---|---|---|
| Genus | *Herpestes* | |
| | *Herpestes ichneumon ichneumon* | Egyptian Mongoose |

Family                      HYAENIDAE: HYAENAS

Genus        *Hyaena*
                *Hyaena hyaena syriaca*                    Syrian Striped Hyaena
                *Hyaena hyaena sultana*                    Arabian Striped Hyaena


Family                      FELIDAE: CATS

Genus        *Felis*
                *Felis silvestris tristrami*                Palestine Wild Cat
                *Felis silvestris iraki*                    Desert Wild Cat
                *Felis margarita*                        Sand Cat
                *Felis chaus furax*                    Palestine Jungle Cat
Genus        *Lynx*
                *Lynx caracal schmitzi*                   Arabian Caracal
Genus        *Panthera*
                *Panthera leo persicus*                   Persian Lion
                *Panthera pardus tulliana*               Syrian Leopard
                *Panthera pardus nimr*                   Arabian Leopard
Genus        *Acinonyx*
                *Acinonyx jubatus venaticus*              Asiatic Cheetah


**Order**                    **UNGULATA: UNGULATES**

Suborder                 PERISSODACTYLA: ODD-TOED UNGULATES


**Order**                    **UNGULATA: UNGULATES**

Family                    PROCAVIIDAE: HYRAXES

Genus        *Procavia*
                *Procavia capensis syriacus*              Syrian Rock Hyrax
                *Procavia capensis sinaiticus*            Sinai Rock Hyrax


Suborder                 ARTIODACTYLA: EVEN-TOED UNGULATES

Family                    SUIDAE: SWINE

Genus        *Sus*
                *Sus scrofa lybicus*                   Palestine Wild Boar

| | | |
|---|---|---|
| Family | | CERVIDAE: DEER |
| Genus | *Dama* | |
| | *Dama mesopotamica* | Mesopotamian Fallow Deer |
| Genus | *Capreolus* | |
| | *Capreolus capreolus coxi* | Kurdish Roe Deer |

| | | |
|---|---|---|
| Family | | BOVIDAE: OXEN, ANTELOPE AND GOATS |
| Subfamily | | ANTELOPINAE: ANTELOPES |
| Genus | *Gazella* | |
| | *Gazella gazella gazella* | Mountain Gazelle |
| | *Gazella gazella cora* | Arabian Gazelle |
| | *Gazella dorcas* | Dorcas Gazelle |

| | | |
|---|---|---|
| Subfamily | | CAPRINAE: GOATS |
| Genus | *Capra* | |
| | *Capra ibex sinaitica* | Sinai Ibex |

**Order**     **LAGOMORPHA: HARES AND RABBITS**

| | | |
|---|---|---|
| Family | | LEPORIDAE: HARES |
| Genus | *Lepus* | |
| | *Lepus capensis syriacus* | Syrian Hare |
| | *Lepus capensis philistinus* ssp. nov. | Philistine Hare |
| | *Lepus capensis sinaiticus* | Sinai Hare |
| | *Lepus capensis arabicus* | Arabian Hare |

**Order**     **RODENTIA: GNAWING ANIMALS**

| | | |
|---|---|---|
| Family | | SCIURIDAE: SQUIRRELS |
| Genus | *Sciurus* | |
| | *Sciurus anomalus syriacus* | Persian Squirrel |

| | | |
|---|---|---|
| Family | | HYSTRICIDAE: PORCUPINES |
| Genus | *Hystrix* | |
| | *Hystrix indica indica* | Indian Crested Porcupine |

| Family | CAPROMYDIAE: COYPUS | |
|---|---|---|
| Genus | *Myocastor* | |
| | *Myocastor coypus* | Coypu |

| Subfamily | DIPODINAE: JERBOAS | |
|---|---|---|
| Genus | *Jaculus* | |
| | *Jaculus orientalis orientalis* | Greater Egyptian Jerboa |
| | *Jaculus jaculus vocator* | Thomas's Lesser Three-toed Jerboa |
| | *Jaculus jaculus macrotarsus* n. status | Wagner's Lesser Three-toed Jerboa |
| | *Jaculus jaculus schlueteri* | Schlueter's Lesser Three-toed Jerboa |

| Family | GLIRIDAE: DORMICE | |
|---|---|---|
| Subfamily | MUSCARDINAE: DORMICE | |
| Genus | *Eliomys* | |
| | *Eliomys quercinus melanurus* n. status | Levant Garden Dormouse |
| | *Eliomys quercinus fuscus* ssp. nov. | Sooty Dormouse |
| Genus | *Dryomys* | |
| | *Dryomys nitedula phrygius* | Forest Dormouse |

| Family | SPALACIDAE: MOLE RATS | |
|---|---|---|
| Genus | *Spalax* | |
| | *Spalax microphthalmus ehrenbergi* n. status | Greater Mole Rat |

| Family | MURIDAE: MICE, RATS AND HAMSTERS | |
|---|---|---|
| Subfamily | MURINAE: MICE AND RATS | |
| Genus | *Apodemus* | |
| | *Apodemus mystacinus mystacinus* | Big Levantine Field Mouse |
| | *Apodemus flavicollis argyropuloi* | Yellow-necked Field Mouse |
| | *Apodemus sylvaticus iconicus* | Common Field Mouse |
| | *Apodemus sylvaticus chorassanicus* | Alpine Field Mouse |
| Genus | *Rattus* | |
| | *Rattus rattus rattus* | Common Rat |
| | *Rattus norvegicus norvegicus* | Brown Rat |
| Genus | *Mus* | |
| | *Mus musculus praetextus* | House Mouse |
| | *Mus macedinicus* | Macedonian House Mouse |

| | | |
|---|---|---|
| Genus | *Acomys* | |
| | *Acomys cahirinus dmidiatus* | Egyptian Spiny Mouse |
| | *Acomys cahirinus homericus* | Southern Spiny Mouse |
| | *Acomys russatus russatus* | Golden Spiny Mouse |
| Genus | *Nesokia* | |
| | *Nesokia indica bacheri* | Short-tailed Bandicoot Rat |

Subfamily            CRICETINAE: HAMSTERS

| | | |
|---|---|---|
| Genus | *Cricetulus* | |
| | *Cricetulus migratorious cinerascens* | Gray Hamster |

Subfamily         GERBILLINAE: GERBILS, JIRDS AND SAND RATS

| | | |
|---|---|---|
| Genus | *Gerbillus* | |
| | *Gerbillus henleyi mariae* | Pygmy Gerbil |
| | *Gerbillus nanus arabium* | Baluchistan Gerbil |
| | *Gerbillus dasyurus dasyurus* | Wagner's Gerbil |
| | *Gerbillus dasyurus leosollicitus* | Lehman's Gerbil |
| | *Gerbillus andersoni bonhotei* | Anderson's Gerbil |
| | *Gerbillus andersoni allenbyi* | Allenby's Gerbil |
| | *Gerbillus gerbillus asyutensis* | Lesser Egyptian Gerbil |
| | *Gerbillus pyramidum floweri* | Great Egyptian Gerbil |
| Genus | *Meriones* | |
| | *Meriones tristrami tristrami* | Tristram's Jird |
| | *Meriones tristrami bodenheimeri* | Syrian Tristram's Jird |
| | *Meriones tristrami deserti* ssp. nov. | Tristram's Desert Jird |
| | *Meriones crassus crassus* | Sundevall's Jird |
| | *Meriones lybicus sacramenti* n. status | Buxton's Jird |
| | *Meriones calurus calurus* | Bushy-tailed Jird |
| Genus | *Psammomys* | |
| | *Psammomys obesus terraesancta* | Fat Sand Rat |

Family              MICROTINAE: VOLES

| | | |
|---|---|---|
| Genus | *Arvicola* | |
| | *Arvicola terrestris hintoni* | Syrian Water Vole |
| Genus | *Microtus* | |
| | *Microtus nivalis hermonis* | Hermon Snow Vole |
| | *Microtus socialis guentheri* | Gunther's Social Vole |

# INSECTIVORES: INSECTIVORA

Insectivores are generally small mammals, characterized by their many small, sharp, pointed teeth; long snout; and plantigrade feet, usually with five toes and all furnished with claws. They are further distinguished by their almost exclusively insect-eating habits. In Israel, they are represented by hedgehogs and shrews. There are no moles in Israel.

Fig. 1. Ethiopian Hedgehog rolled up.

# HEDGEHOGS: ERINACEIDAE

Hedgehogs are fairly small in size and distinguished by a coat of short, dense spines that covers the back and sides of the body. The tail is short and scarcely visible. The hedgehog is noted for its peculiar ability to roll up into a ball, exposing only its spiny back (Fig. 1). Their diet is mostly invertebrates, small animals and almost anything, including carrion. It can kill a viper although it is not immune to its venom. Confined to the Old World, it occurs in both temperate and arid regions. In Israel, hedgehogs are represented by three species.

## Eastern European Hedgehog     Plate 1

*Erinaceus europaeus concolor* Martin, 1838

Kipod Matsui (Hebrew)

Qunfidha, Kanfood, or Kunfuth (Arabic)

**Description**: Head and body 197.5–243.4 mm; ear 23–31.2 mm.; hind foot 37–45.5 mm; tail 18–38.6 mm.

The Eastern European Hedgehog is easily distinguished from other hedgehogs by its overall brown color, sometimes blackish-brown; with a dark gray-brown belly, small ears and brownish face. The carapace has no dark median stripe. This race is characterized by a white chest. The legs are brown, and the short, naked tail is usually not discernible. In paler forms, the brown of the face is mixed with white hairs, particularly between the ears and the forelegs, on the throat and behind the eyes, where it becomes almost white.

**Habits**: Mainly nocturnal, it is sometimes seen during the day. Its voice is a hog-like grunt. Heterothermic, it can regulate its body temperature physiologically. It hibernates for short periods. Gestation is 31–35 days, and the litter is 2–10.

**Distribution**: Transcaucasia, Asia Minor, Lebanon and Syria. In Israel, it is common in the northern half of the country, in the Judean hills (around Jerusalem) and along the coastal plain, as far south as Ruhama.

**Texomatic Remarks**: *E. e. concolor* is considered by some as a full species (Macdonald and Barrett, 1993). It is distinguished by its white chest patch and the postero-dorsal processes of the maxillae reach behind the lachrymal foramena. However, it replaces *E. e. europaeus* geographically and hybridizes with it. Genetically, it is questionably a distinct species.

## Egyptian Long-eared Hedgehog     Plate 2

*Hemiechinus auritus aegyptius* Fischer, 1829

Kipod Holot (Hebrew)

Qunfidha, Kanfood or Kunfuth (Arabic)

**Description**: Head and body 142–202 mm;

ear 30–43 mm.; hind foot 28–35 mm.; tail 11–37 mm.

The Egyptian subspecies of the Eurasian Long-eared Hedgehog is the smallest of the three hedgehogs found in Israel, with relatively longer legs. It is easily recognized by its small, pale appearance; large, roundish ears; whitish face, sometimes tinged with buff; a carapace without a dark median line; and whitish legs and underparts. The spines are short and fine, not parted on the crown. The juvenile has a dark muzzle.

**Habits:** It is nocturnal, spending the day in a burrow. It hibernates for 5–40 days. Gestation is 36–37 days, and the litter 1–5.

**Habitat:** Semi-arid, more sandy places with vegetation.

**Distribution:** Egypt and Cyrenaica. In Israel, it is common in the southern part of the coastal plain as far north as Caesarea, and south to the northern Negev Desert (around Beersheba).

**Taxonomic Remarks:** Two subspecies of the Long-eared Hedgehog, *H. a. aegyptius* and *H. a. calligoni* (Satunin), 1901, are said to intergrade in Israel (Harrison and Bates, 1991). The eastern subspecies, *H. a. calligon*, usually has a pure white abdomen, which may be yellowish or in the western subspecies, *H. a. aegyptius* washed with brown. However, this is variable, and *H. a. calligoni* does not differ from *H. a. aegyptius* in any significant character. It is doubtful whether the subspecies *H. a. calligoni* is valid.

**Ethiopian Hedgehog, Arabian "Porcupine"**              Plate 2, Figure 1

*Paraechinus aethiopicus pectoralis* Anderson and de Winton, 1901

Kipod Midbar (Hebrew)

Qunfidha, Kanfood, or Kunfuth (Arabic)

**Description:** Head and body 120–212 mm.; ear 34–48 mm.; hind foot 26–38 mm.; tail 13–31 mm.

The size of the Ethiopian Hedgehog is between that of the Egyptian Long-eared and the Eastern European Hedgehogs. The upper parts vary from light brown to blackish-brown, particularly on the back. It is easily distinguished by the distinctly parted spines on the crown; a bicolored face, strongly marked with blackish-brown; and a white forehead band or all-brown face. The carapace usually has a dark median stripe. The ears are large; the legs brown; and there is a distinct white area between the ears and forelegs. The rest of the underparts vary in amounts of brown or white.

**Habits:** It is nocturnal and has extraordinary hearing; otherwise little is known. It practices estivation (dormancy in response to heat, drought and shortage of food). Gestation is about 35 days, and the litter is 1–5.

**Habitat:** Desert hills, wadis and plains with some vegetation.

**Distribution:** Africa, Sinai and Arabia. In Israel, it is known from the Negev Desert (south of Beersheba) and reaches the Jordan Valley.

# SHREWS: SORICIDAE

The shrew is among the smallest and most inconspicuous of mammals, even where it is common. Superficially, it resembles a mouse, but may be distinguished by its protracted, mobile snout; tiny eyes; short, external ears which are sometimes concealed by short, velvety fur; and by its teeth which have pointed, not chisel-shaped, incisors. The teeth have chestnut-colored tips or are all white. In addition, it has five clawed toes on each foot, whereas mice have only four clawed toes on their forefeet. Widespread in distribution, it is found in both the New and Old Worlds, where it generally prefers moist regions, but it occurs in the desert as well. The shrew is an extremely active and rapacious insectivore. It forages under logs, fallen leaves, plant debris and narrow crevices beneath rocks. The White-toothed Shrew engages in "caravanning", whereby after the young leave the nest, they hang onto each other and their mother in a procession (Fig. 2). In Israel, shrews are represented by five species.

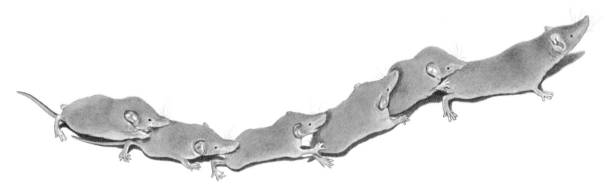

Fig. 2. When the young Greater White-toothed Shrews leave the nest they hold on to the mother and each other in a procession called "caravanning."

**Pygmy, Etruscan or Lesser Shrew**                    Plate 1

*Suncus etruscus etruscus* Savi, 1822
Hadaf Zair (Hebrew)
Far el Kla (Arabic)

**Description**: Head and body 37.1–54 mm.; ear 3–6 mm.; hind foot 6.3–8 mm.; tail 21–31.3 mm.

The Etruscan Shrew weighs two grams, and is the smallest living mammal in the world. The upper parts are dark brownish-gray, with a slight sheen. The underparts are a lighter gray or dull white. Its snout is somewhat elongated, and the relatively long tail is thickly clad with short hairs and sparsely with long hairs.

**Habits**: Mostly nocturnal, it is fairly tame and

may allow human handling, although it is aggressive towards its own kind. Its nest is similar to that of the White-toothed Shrew. Gestation is 27–28 days, and the litter is 2–5.

**Habitat**: Meadows, hedgerow banks and gardens, sometimes entering houses.

**Distribution**: Widespread from southern Europe, the Mediterranean region, the Caucasus, southern Russia, Turkestan and Iran, Arabia, northern Nigeria, French Guinea, North Africa, East and South Africa, India, Ceylon and Malaya. In Israel, it has been recorded from the northern part of the country as far south as Mar Saba near the Dead Sea.

### Bicolor White-toothed Shrew     Plate 1, Figure 4

*Crocidura leucodon judaica* Thomas, 1919
Hadaf Levan Shenayim (Hebrew)
Far el Kla (Arabic)

**Description**: Head and body 50.3–78.7 mm.; ear 4–9.5 mm.; hind foot 11.8–13 mm.; tail 35–40.4 mm.

The Bicolor White-toothed Shrew is of medium size, with dull brown or reddish-brown upper parts, which are distinctly demarcated from the whitish underparts and visible from the side. These extend to the eyes and angle of the mouth. The tail is relatively short, usually less than half the length of the head and body, and bicolored, dark above and paler below. The feet are whitish with white claws. The dentition can be distinguished by $I^2$

which equals the lengths of $I^3$ and C combined, and $I^3$ is the smallest of the three unicuspids (Fig. 3).

**Habits**: Mostly nocturnal, it inhabits low undergrowth, edges of woods and gardens. The nest of grass is made in sheltered undergrowth. Gestation is 28–33 days, and the litter is 2–10.

**Habitat**: Plains and mountains up to 3,600 m. in dense vegetation.

**Distribution**: From western Europe to central and southern Russia, eastern Turkestan, Asia Minor, northern Iran and the Arabian peninsula. In Israel, it is found in the northern and central part of the country, Mount Hermon (at 1,550 m.) and the Golan Heights.

### Thomas's White-toothed Shrew     Plate 2

*Crocidura lasia* Thomas, 1907
Hadaf (Hebrew)
Far el Kla (Arabic)

**Description**: Head and body 72–92 mm.; ear 8.5–12 mm.; hind foot 12.5–13.4 mm.; tail 38–45 mm.

Thomas's White-toothed Shrew is similar to the Bicolor White-toothed Shrew, only slightly larger, on average, with its warm brown upper parts, not sharply demarcated from the whitish underparts. The feet are less robust and not as white, and the tail is longer, usually more than half the length of the head and body.

**Habits**: It is active at night; otherwise little is known.

**Habitat**: Rocky mountains with low vegetation.

**Distribution**: Turkey, Caucasia, Iran and Lebanon. In Israel, it has been recorded once from Mount Hermon at 1,200 m. above sea level.

**Taxonomic Remarks**: Thomas (1906) described *C. leucodon lasia* from northeastern Turkey. He later decided it was a distinct species, and it became *C. lasia* (Thomas, 1907). It was then assigned as a subspecies of *C. lasiura* (Bobrinksy et al., 1944), followed by Ellerman and Morrison-Scott (1951) and Harrison (1964). Lay (1967) showed that *C. lasiura* is distinct from *C. lasiura lasia*, and that *C. leucodon lasia* was, in fact, valid. Since it is unlikely that two subspecies of *C. leucodon*, *C. l. judaica* and *C. l. lasia*, would be sympatric on Mount Hermon, and that *C. lasia* can be distinguished from *C. leucodon*, this author agrees with Thomas and Ognev (1928) in recognizing *C. lasia* as a distinct species.

## Great White-toothed Shrew

Plate 1, Figure 2,3

*Crocidura russula monacha* Thomas, 1906
Hadaf Matsui (Hebrew)
Far el Kla (Arabic)

**Description**: Head and body 58.5–81.8 mm.; ear 4–9.8 mm.; hind foot 10.3–13.6 mm.; tail 40–48.8 mm.

This medium-sized shrew is very variable in color, ranging from reddish to grayish-brown upper parts, which are not sharply demarcated from the lighter gray-brown or yellow-gray underparts. The long tail is characteristic, more than half the length of the head and body. On the Golan Heights (Masada Forest), it is sooty-gray, the underparts not tipped with whitish-buff as in the nominate form. The juvenile is gray and can be recognized by its adolescent proportions. For dentitions see *C. suaveoleus* (Fig. 3).

**Habits**: Mostly nocturnal, it builds its nest of grass lined with leaves. In the winter, communal nests have up to six shrews sleeping together in a hollow in the ground, or in an open saucer shape. Gestation is 28–33 days, and the litter is 2–10.

**Habitat**: Forests, open fields and gardens among leaves and rocks, often near water.

**Distribution**: Widespread in western Europe, southern Russia, Transcaucasia, southern Turkestan, Armenia, Cyprus, Asia Minor, Iran, Afghanistan, east to Kashmir, China, Japan and North Africa, from Morocco to Tunisia, but not Egypt. The subspecies *C. r. monacha* is known from Lebanon and Israel, where it is the most common shrew from Mount Hermon and the Golan Heights as far south as Eilat.

**Taxonomic Remarks**: There has been much controversy over the presence of *C. russula* in Israel. Catezeflis et al. (1985) claim that all the specimens attributed to *C. russula* are really *C. suaveolens*. Shalmon (1993) and Mendelssohn and Yom-Tov (1999) agree. Several authorities, however, recognize that *C. russula* and *C.*

Fig. 3. Anterior teeth of white-toothed shrews. Left to right: *Crocidura leucodon*, *Crocidura russula*, *Crocidura suaveolens*

*suaveolens* are distinct and recognizable species (Van den Brink, 1956; Harrison, 1964; Lay, 1967; Harrison and Bates, 1991; and MacDonald and Barret, 1993). The adult *C. suaveolens* is usually smaller than *C. russula* in total length and greatest length of the skull, although there is some overlap. In their description of *C. suaveolens*, Mendelssohn and Yom-Tov (1991) fail to mention the most diagnostic feature of *C. suaveolens* — the sharp demarcation between the white ventral streak and dark brown upper parts. This character is lacking in most specimens of *Crocidura* in Israel, as in *C. russula*, which is characterized by a grayish belly, not sharply delimited from the color of the upper parts. The most common shrew in Israel is, in fact, *C. russula*, and not *C. suaveolens*.

## Lesser White-toothed Shrew  Plate 1, Figure 3

*Crocidura suaveolens portali* Pallas, 1811

Hadaf Katan (Hebrew)

Far el kla (Arabic)

**Description:** Head and body 57–67 mm.; ear 6–9 mm.; hind foot 10–12.5 mm.; tail 35–41.8 mm.

Similar to the Great White-toothed Shrew, the Lesser White-toothed Shrew can be distinguished by its smaller size and relatively long tail. It is gray, tinged with brown above, paler below, and characterized by a distinct narrow white ventral streak the length of the abdomen, not visible from the sides. The feet are whitish.

*Crocidura suaveolens* can be distinguished from *C. russula* by its dentition (Fig. 2):

| *C. suaveolens* | *C. russula* |
|---|---|
| $I^1$ not prominent | $I^1$ prominent, hook-shaped |
| $I^2$ length exceeds combined lengths of $I^3$ and C | $I^2$ length less than combined lengths of $I^3$ and C |
| $I^3$ smallest of 3 unicuspids, smaller than C | $I^3$ subequal with C |

**Habits:** A rare and little known species, its nest is of soft vegetation in sheltered places. Gestation is 28 days, and there are up to four litters per year, each with 1–6 young. It feeds on insects, arthropods, molluscs and worms. In winter, it sometimes enters human habitation and eats hibernating spiders.

**Habitat**: Forest edges, meadows and gardens. In the desert, it was found in a date plantation.

**Distribution**: Europe and Asia. The subspecies *C. suaveolens portali* is known from northern Iraq and Egypt (southern Sinai). In Israel, it has been found in the Huleh Valley (Dan), the coastal plain (Ramleh) and in the Arava Valley (Ein Yahav).

## Ramon White-toothed Shrew          Plate 1

*Crocidura ramona* Ivanitskaya Shenbrot and Nevo, 1996
Hadaf Ramon (Hebrew)
Far el kla (Arabic)

**Description**: Head and body 58–63 mm.; ear 7.5–8 mm.; hind foot 10–11 mm.; tail 42–43 mm.

The Ramon White-toothed Shrew is relatively small with a flattened head. The upper parts are a light slaty-gray, with white tips that give a silvery appearance. The underparts are silvery-white, clearly demarcated but not very sharp. The tail is indistinctly bicolored, with the dorsal surface lighter than the back.

**Habits**: Unknown.

**Habitat**: Rocky desert highlands, in wadis with Tamarix and Retama and dense Atriplex vegetation.

**Distribution**: Presently known only from southern Israel, at the northern edge of the Judean Desert and the Negev Desert highlands (Mizpe Ramon, Sde Boqer and Sartaba).

**Taxonomic Remarks**: *Crocidura ramona* may be a synonym of Tristram's Shrew, *Suncus tristrami*, Bodenheimer, 1935, a small, gray, *silvery* shining species found by Tristram in the Negev Desert (Bodenheimer, 1935).

## Pale Gray Shrew, East Persian          Plate 2
## White-toothed Shrew

*Crocidura pergrisea* Miller, 1913
Hadaf Afor Bahir (Hebrew)
Fal el kla (Arabic)

**Description**: Head and body 53–72 mm.; ear 7.5 mm.; hind foot 11–14 mm.; tail 36–54 mm.

The Pale Gray Shrew is of medium size, slightly larger than *C. suaveolens*, and generally with a longer tail. The upper parts are gray tipped with brown. The underparts are a dull white, indistinctly demarcated along the flanks. The feet are dusky white, faintly darker on the outer two toes of the hind feet. The tail is grayish-brown above and paler with whitish hairs below.

**Habits**: It inhabits mountains at high elevations (3,115 m. in Baltistan, Kashmir). Otherwise, very little is known.

**Habitat**: Rocky walls with cracks and holes.

**Distribution**: Baltisan (Kashmir), eastern Iran. In Israel, it has been recorded once on Mount Hermon at 2,000 m.

# BATS: CHIROPTERA

Bats are the only mammals capable of true flight, as their forelimbs have been modified into wings. There are two basic wing shapes, long and narrow for fast flight, and short and broad for slow maneuverability. An interfemoral membrane extends between the elongated phalanges and the forearm to the side of the body and hind leg and again between the hind leg and tail, varying in development, or absent. When present, the free edge of this membrane is supported by a cartilaginous calcar, or spur, connected to the inner side of the foot. Some bats have a free edge beyond the calcar. Nearly all bats have small eyes and poor vision. They use echolocation to avoid objects and locate prey. Most species of bats are social and collect by the thousands, but some are solitary. Bats spend half their lives hanging upside-down. Numerous species migrate in the winter when insects are scarce, or hibernate. They are mainly nocturnal and crepuscular. Many species are polygamous. Homosexuality is common, even between different species. Adverse weather and lack of food produce torpidity, which lengthens the gestation period. They give birth hanging upside-down. Bats are among the few mammals which do not provide homes (dens or nests) in which to raise their young.

Bats comprise the second largest order of mammals, after rodents. They are divided into two suborders, frugivorous and insectivorous bats. Bats are more or less cosmoplitan. In Israel, there is a single family of frugivorous bats and seven families of insectivorous bats, well-represented by 32 species, almost as many as in all of Europe.

## Names of parts of a bat

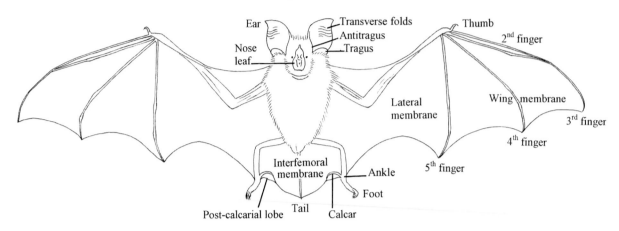

## Parts of the nose leaf

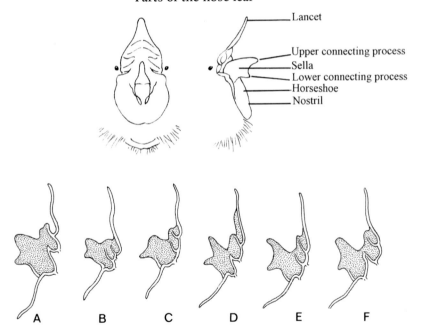

Fig. 4. Anatomy of the bat
The distinctive profile of the sela in the nose-leafs of horseshoe bats (left lateral view).
A.  Greater Horseshoe Bat, *Rhinolophus f. ferrum-equinum*
B.  Desert Horseshoe Bat, *Rhinolophus c. clivosus*
C.  Lesser Horseshoe Bat, *Rhinolophus hipposideros minimus*
D.  Peter's Horseshoe Bat, *Rhinolophus blasii*
E.  East Mediterranean Horseshoe Bat, *Rhinolophus euryale judaicus*
F.  Mehely's Horseshoe Bat, *Rhinolophus mehelyi*

# FLYING "FOXES": MEGACHIROPTERA

The frugivorous bats are characterized by their large size. They are also called Flying "Foxes," so named for their fox-like heads with a moderately long muzzle and large pointy ears. With their large eyes they appear to guide themselves visually in flight. The locate food, which consists of fruit, by smell. The fruit pulp is crushed, most of which they spit out after swallowing the juice. Fruit that is dropped is not retrieved. In Israel they are represented by a single species.

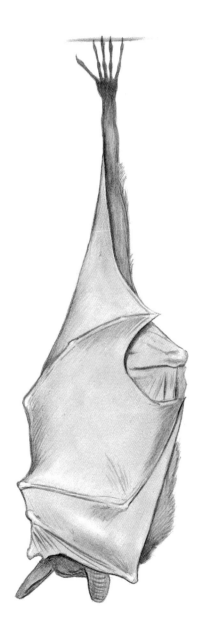

Fig. 5. Egyptian Fruit Bat asleep

45

# FRUIT BATS: PTEROPODIDAE

The family of Fruit Bats, or true Flying "Foxes", comprises two families, the Pteropodidae and the Megaloglossidae. The Pteropodidae is confined to the tropics and subtropics of the Old World. Fruit Bats feed mainly on fruit juices and nectar. They are represented in Israel by a single species.

**Egyptian Fruit Bat,**       Plate 3, Figure 5
**Flying "Fox"**

*Rousettus aegyptiacus aegyptiacus* E. Geoffroy, 1810

Atalef Perot (Hebrew)

Khaffash, Wat-Wat (Arabic)

**Description:** Head and body 50–110.7 mm., male larger than female; ear 18.6–25 mm.; forearm 84.1–100 mm.; wingspread 35.6–53.3 mm.; tail 8.3–19 mm.

The Egyptian Fruit Bat is the largest bat found in Israel. It is further distinguished by its fox-like head, which has a projecting muzzle; simple, unspecialized nose pad; large eyes; relatively small ears without any tragus, the sides of which are united at the base forming a complete ring. The wings differ from all other bats in that they have three, rather than one, or rarely two joints in the second or index finger, and bear a claw found only on the thumb of other bats. The very short tail is entirely connected to the narrow interfemoral membrane. The general color is yellowish or sooty-brown. The juvenile is grayer.

**Habits:** Gregarious, it roosts in large noisy colonies. Unlike other large bats, it uses its eyes and clicks its tongue for acoustical orientation. It is helpless in total darkness. Its flight is steady and direct, and it may fly up to 15 km. for foraging. It is frugivorous, eating dates, carob, citrus and chinaberry.

**Habitat:** Unlike other fruit bats that roost only in trees, this species inhabits caves and deserted buildings, not far from water.

**Distribution:** Cyprus, Syria, Lebanon, Egypt, Ethiopian Africa in part, south to Angola. In Israel, it is common in the coastal plain, on Mount Carmel, in the Huleh and Jordan Valleys, near the Dead Sea (Ein Gedi), but rarely in the hills. In Eilat, it probably represents the Arabian subspecies, *R. aegyptiacus arabicus* Anderson and de Winton, 1902, which is usually smaller, with a more pointed ear tip.

46

# INSECTIVOROUS BATS: MICROCHIROPTERA

The insectivorous bats include many families. They inhabit most of the temperate and tropical regions of the world. Insectivorous bats are relatively small in size and constitute the major part of the bat population. Their flight is very maneuverable. Echolocation allows them to catch unseen insects and to avoid obstructions in their path. Their food, which consists mainly of insects, is usually caught on the wing. Heterothermic, they use food at a lowered metabolic output or hibernate. Bats represent virtually one third of all species in Israel.

Fig. 6. Egyptian Slit-faced Bat at rest

# MOUSE–TAILED BATS: RHINOPOMATIDAE

The Mouse-tailed Bat is characterized by a short, triangular nose-leaf immediately above the nostrils; fairly large, united ears with a distinct, fringed tragus; two phalanges in the second digit of the manus; and an extremely long tail which projects freely beyond the short, interfemoral membrane. It does not hibernate but acquires fat and becomes torpid during cold, dry weather. This bat occurs in hot, arid areas. It can conserve water by concentrating urine.

The Mouse-tailed Bat is found in subtropical and tropical parts of northern and northeast Africa, Arabia and India to the Malayan Peninsula. In Israel, it is represented by two species.

## Larger Mouse-tailed, Great Rat-tailed or Long-tailed Bat

Plate 4

*Rhinopoma microphyllum microphyllum* Bruennich, 1782

Atalef Yaznuv Gadol (Hebrew)

Khaffash (Arabic)

**Description:** Head and body 132–133 mm.; ear 19.4–21.8 mm.; forearm 64–71.7 mm.; wingspread 31.8 mm.; tail 55–62.5 mm.

The Larger Mouse-tailed Bat is normally distinguished by its very long, slender tail which projects freely well beyond the margin of the short interfemoral membrane. Often, however, the tail is absent. There is no distinct nose-leaf, but a fleshy prominence lies above the nose. The upper parts are uniformly ashy-brown and dusky-white below. The tail is generally shorter than the forearm.

**Habits:** It acquires large fat deposits at the base of the tail before hibernation.

**Habitat:** Caves, clefts in rocks, old ruins, often not far from water.

**Distribution:** Northeast Africa, Egypt and Arabia, through India and Burma. In Israel, it is fairly common in the Huleh Valley, and rarer in the hills, Jordan Valley, plain of Jericho and the Dead Sea basin.

## Lesser Mouse-tailed Bat

Plate 4

*Rhinopoma hardwickei cystops* Thomas, 1903

Yaznuv Katan (Hebrew)

Khaffash (Arabic)

**Description:** Head and body 50–75 mm.; ear 15–20.8 mm.; forearm 47.8–59.3 mm.; wingspread 25.5–29.2 mm.; tail 39.6–79 mm.

The Lesser Mouse-tailed Bat is distinguished by its generally small size and upper parts of light brownish-gray color, with a paler and relatively long tail. The free portion of the tail is usually longer than the forearm.

**Habits**: It is nocturnal and it hibernates in the winter. The juvenile hangs on to prepubic teats at birth, for security as well as nourishment.

**Habitat**: Ruins, dry caves and rock crevices on mountains as high as 1,220 m. above sea level.

**Distribution**: North and East Africa, the Asben region, Tunisia, Egypt, Sudan, southern to northern Kenya, Arabia, Iran, Afghanistan, India, Burma and lower Siam. The subspecies *R. h. cystops* occurs in Iraq and Jordan. In Israel, it is found in the Huleh Valley (Dan), western Galilee (Hanita and Rosh Hanikra), the western shore of the Sea of Galilee (Wadi Amud), the western shore of the Dead Sea (Ein Gedi) and the Arava Valley (Ein Yahav).

# OLD WORLD SHEATH-TAILED BATS: EMBALLONURIDAE

The Sheath-tailed or Tomb Bat is characterized by its triangular, pointed snout and broad, separated ears with a short, blunt, expanded tragus. The tail perforates the interfemoral membrane about the middle, leaving the last three or four vertebrae free above the membrane, but it can be retracted into the membrane and does not project beyond the hind edge.

Sheath-tailed Bats are a subtropical or tropical species found in the Ethiopian, Oriental and Australian regions and the southernmost part of the Palearctic Region. In Israel, they are represented by two species.

## Naked-bellied Tomb Bat     Plate 3

*Taphozous nudiventris nudiventris*
Cretzschmar, 1830 *vel* 1831
Ashman Gadol (Hebrew)
Khaffash (Arabic)

**Description**: Head and body 69–97.5 mm.; ear 18–24 mm.; forearm 64.9–79.2 mm.; wingspread about 220 mm.; tail 19–41mm.

The Naked-bellied Tomb Bat is large and stocky, with a large wingspread; tawny or grayish-brown upper parts and grayish-white below. The lower abdomen and thighs develop fat and are naked. It has no nose-leaf, and the ears are comparatively simple. The male has a glandular patch under the chin. The short tail perforates the interfemoral membrane near the center and it either projects or can be withdrawn.

**Habits**: It is nocturnal and acquires a large deposit of fat at the base of the tail before it hibernates. Gestation is 63 days with a single young.

**Habitat**: Hollow trees, caves, crevices, tombs and buildings.

**Distribution**: Egypt, south to northern Kenya, Congo, southern Iran, Arabia, east to India, Burma and Malay States. The subspecies *T. n. nudiventris* is known from Arabia. In Israel, it occurs in the Huleh Valley (Dan), Galilee (Wadi Amud), southeast of Haifa, near the Sea of Galilee and the Dead Sea.

## Tomb Bat, Egyptian Tomb Bat     Plate 3

*Taphozous perforatus haedinus* Thomas, 1925
Ashman Katan (Hebrew)
Khaffash (Arabic)

**Description**: Head and body 56.2–85 mm.; ear 14–16 mm.; forearm 58–66 mm.; tail 21–30 mm.

The Tomb Bat is of medium size, and its fur reaches the base of the tail membrane, both above and below. The upper parts are smoky-brown, paler on the throat, with a gray or grayish-brown belly. The ear is tall, somewhat narrow and shorter than the head. The anitragal lobe is barely developed. The tragus is narrow at its base and club-shaped. The basal lobule is almost completely absent. The calcar supports the outer half of the interfemoral membrane with no post-calcareal lobe. The tail emerges about midway on the interfemoral membrane.

**Habits:** It is nocturnal and inhabits crevices in cliffs, tombs and buildings. Its flight is high, strong and fast. Otherwise, little is known.

**Habitat:** Rocky desert, tombs, small caverns, sea caves, cultivated lands and houses.

**Distribution:** Southwestern Arabia, Egypt, southward and westward in Africa and India. In Israel, the subspecies *T. p. haedinus* has been recorded from west of the Sea of Galilee, the western shore of the Dead Sea, and the northern Negev Desert.

# SLIT-FACED BATS: NYCTERIDAE

The Slit-faced Bat is characterized by the muzzle which is transversed by a longitudinal furrow, posteriorly expanded into a deep pit on the forehead, hidden by nose-leafs. The large ears are united at the base, with a short, more or less rounded tragus. The second digit of the manus is represented only by the metacarpal. The tail is long and terminates in a T-shaped vertebra. Slit-faced Bats are found in Africa, Arabia, the Malay Peninsula and Java. In Israel, they are represented by a single species.

**Egyptian Slit-faced Bat**        Plate 5, Figure 6

*Nycteris thebaica thebaica* Geoffroy, 1818

Leylan (Hebrew)

Khaffash (Arabic)

**Description**: Head and body 45–75 mm.; ear 28–34 mm.; forearm about 36–60 mm.; wingspread 253–280 mm.; tail 43–75 mm.

The Slit-faced Bat is easily distinguished by its small size; minute eyes; longitudinal furrow that traverses the muzzle, concealed by small, indistinct nose-leafs; enormous, rounded ears, much longer than the head; and short, rounded tragus, 3 mm. in height. The upper parts are a pale smoky-fawn, paler below.

**Habits**: Its flight is in short intervals and slow, often near the ground. It begins long after dusk and continues until well before dawn. The gestation is about 155 days, and there is only a single young. Its diet consists largely of arthropods, picked off the ground or from vegetation. Otherwise, little is known.

**Habitat**: Savannah, desert, ruins and houses, hollow trees, rock fissures and the burrows of porcupines. It has been found up to 1,830 m. above sea level.

**Distribution**: Greece (Corfu), Egypt (Sinai), Sudan, Kenya, Angola and Arabia. In Israel, it has been recorded in the Jordan Valley (Beit Shan) and the Arava Valley (Ein Yahav).

# HORSESHOE BATS: RHINOLOPHIDAE

The horseshoe bat is characterized by the naked, leaf-like structure, consisting of three appendages, surrounding the nose. The broad forepart, which partly covers the muzzle, appears in the shape of a horseshoe, so that it seems to be smiling. Protruding above and behind the nostrils is a central structure called the sella, and above and behind the sella is an erect, tapering part called the lancet. The horseshoe bat is further distinguished by its ears, which are fairly large, pointed, usually separated and devoid of any tragus. The lower border of the conch, however, is expanded so as to form a broad, shallow antitragus. The metacarpal of the fourth digit is longer than that of the second digit. The horeshoe bat has four mammae, two are false and are used by the young for grasping.

Horseshoe bats have a wide environmental tolerance in temperate, subtropical, tropical and desert regions of the eastern hemisphere. In Israel, they are represented by six species.

## Greater Horseshoe Bat — Plate 6, Figure 44

*Rhinolophus ferrum-equineum ferrum-equineum* Schreber, 1774

Parsaf Gadol (Hebrew)

Khaffash (Arabic)

**Description**: Head and body 49.8–76 mm., female usually larger than male; ear 19.3–27 mm.; forearm 53.3–60 mm.; wingspread 317–358 mm.; tail 31–43 mm.

The Greater Horseshoe Bat is distinguished by its large size and distinct horseshoe-shaped nose-leaf. The upper connecting process of the sella is short and broad, much larger than the pointed lower connecting process (Fig. 4). The ears are moderately large, pointed, longer than they are broad and without a tragus. The upper parts are a warm tawny-gray, or brightly tinged with reddish, browner in the female, and often very pale, almost white in the male. The underparts are paler. The first coat of the young is a pale gray. The wings are relatively broad, making it appear larger than it really is.

**Habits**: Its flight begins about an hour after sunset and continues at intervals until just before sunrise. It is short, usually low, up to about 10.6 meters, slow and graceful, sailing, fluttering and buoyant like a butterfly, often interrupted for feeding. When alighting to rest, it turns a complete somersault and hangs head downward with the tail cocked over the back. It hibernates with its body enshrouded by the wing membranes.

**Habitat**: Caves, catacombs, and old buildings.

**Distribution**: Widespread from Europe through Asia, north of the Himalaya Mountains, as far east as Japan, south through Asia Minor, Syria and Africa. In Israel, it is common throughout the Mediterranean coastal region, around the Sea of Galilee and the Judean hills (Jerusalem).

## Desert Horseshoe Bat          Plate 6, Figure 44

*Rhinolophus clivosus clivosus* Cretzschmar, 1828

Parsaf HaNegev (Hebrew)

Khaffash (Arabic)

**Description**: Head and body 48.3–54.8 mm.; ear 18–21.8 mm.; forearm 45.6–49.9 mm.; tail 19.6–33 mm.

The Desert Horseshoe Bat is of medium size, light gray-brown above, lighter below, or pale buff-brown. It may be distinguished by the ears, which are considerably shorter than the head and acutely pointed. The inner margin of the ear conch is convex throughout, but less convex above; the tip is not attenuated, with the antitragus separated by an obtuse notch, short and evenly convex, and by the form of the upper margin of the same connecting process. The nose-leaf does not conceal the muzzle and is divided by a simple notch in front. In the sella, the upper connecting process is small, short and bluntly pointed, barely larger than the lower connecting process (Fig. 4). The horizontal plate of the sella has semi-circular sides which are much broader than the vertical process. This is slightly narrower in the middle and similar to that of the Greater Horseshoe Bat. The summit of the nose-leaf is rounded above, with the crest of the connecting process much higher and forming an acute, thickened projection. The terminal leaf is rather short and acute. The wings arise from the tibiae slightly above the ankles, and the last small, caudal vertebrae projects.

**Habits**: It is either solitary or gregarious and inhabits caves and cellars. Otherwise, little is known.

**Habitat**: Desert.

**Distribution**: Red Sea coasts of Arabia, African coast of Gulf of Aden, southern Arabia, Eritrea, Egypt, the Sinai and Sahara Deserts. In Israel, it may occur in the southern coastal plain, and northern Negev Desert.

## Lesser Horseshoe Bat          Plate 6, Figure 44

*Rhinolphus hipposideros minimus* Heuglin, 1861

Parsaf Gamadi (Hebrew)

Khaffash (Arabic)

**Description**: Head and body 33–40 mm.; ear 12.7–15.2 mm.; forearm 26.9–37.6mm.; wingspread 153–222 mm.; tail 22–29 mm.

The Lesser Horseshoe Bat is distinguished by its small-sized nose-leaf and proportionately larger and more pointed ears. The male is a pale sandy color above, while the female is browner, and both have lighter underparts. The first coat of the young is pale gray. It has a fan-shaped posterior nose-leaf with various small processes on the upper edge. The upper connecting process of the sella is low and blunt, shorter than the prominent lower connecting process (Fig. 4). The toes of the hind feet have only two joints.

**Habits:** It begins flying at dusk and continues throughout the night. Its flight is fluttering and irregular though powerful and sustained. It generally flies higher than the Greater Horseshoe Bat and away from trees. It is colonial but may be solitary in the winter.

**Habitat:** Hills, partly wooded, caves, buildings, cultivated lands and the desert, in limestone caves and old ruins, sometimes entering houses.

**Distribution:** Western Europe through to southwestern Russia, Asia Minor, Arabia, Iran, east to Kashmir and North Africa, Sudan and Ethiopia. The subspecies *R. h. minimus* occurs in Lebanon and Arabia. In Israel, it ranges from the Huleh Valley, around the Sea of Galilee, Mount Carmel, Judean hills (Jerusalem), the coastal plain, Negev Desert and Arava Valley (Ein Yahav).

## Peter's Horseshoe Bat, Blasius's Horseshoe Bat

Plate 6, Figure 44

*Rhinolophus blasii* Peters, 1866
Parsaf Matsui (Hebrew)
Khaffash (Arabic)

**Description:** Head and body 40.4–50.1 mm.; ear 17.2–20 mm.; forearm 45–47.2mm.; wingspread about 178 mm.; tail 20.4–20.5 mm.

This medium-sized horseshoe bat is reddish-brown above and light grayish below. It rarely has faint dark "spectacles" around the eyes. It is distinguished by the sella, which has cuneate sides. The vertical process of the sella is short and becomes abruptly narrow half-way up. The upper connecting process of the sella is straight, bluntly pointed and distinctly longer than the lower connecting process (Fig. 4). It terminates in a subacute point, the upper half presenting an anterior triangular surface. The upper margin of the posterior connecting process forms a long, narrow, acutely-pointed projection extending high above the summit of the vertical process of the sella. The terminal nose-leaf is moderately large, and the sides of the terminal triangular process are slightly concave. The lower lip has a single groove. The ears are shorter than the head, and acutely pointed, but much less than those of the Greater Horseshoe Bat. The antitragus is separated from the outer margin by a very shallow emargination. The nose-leaf is a small, horizontal, horseshoe-shaped membrane which leaves the sides of the muzzle uncovered. The wings arise from the ankles, with the interfemoral membrane square behind, and only the extreme tip of the tail projecting.

**Habits:** Colonial, it spends the day sleeping and preening. It hibernates in crevices.

**Habitat:** Caves, in mountains up to 1,200 m.

**Distribution:** Southern Europe, Cyprus, Transcaucasia, southwest Russian Turkestan, Iran, Asia Minor and North Africa. In Israel, it is found from the coastal plain (Herzlia) to the Judean hills (Jerusalem — the Cave of Adullam).

## East Mediterranean Horseshoe Bat

Plate 6, Figure 44

*Rhinolphus euryale judaicus* Anderson and Matschie, 1904
Parsaf Bahir (Hebrew)
Khaffash (Arabic)

**Description**: Head and body 59.8–85 mm.; ear 18–24 mm.; forearm 45–49.7 mm.; wingspread about 255 mm.; tail 20.2–31 mm.

The East Mediterranean Horseshoe Bat is of medium size, with pale brown upper parts and nearly pure white underparts. It is distinguished by the shape of the nose-leaf in which the lancet is the most definitive external character. It is essentially triangular, tapering gradually towards the tip, but may have more- or less-pronounced lateral cavities, and this may cause confusion. The upper connecting process of the sella is pointed, slightly bent downwards, and longer than the lower connecting process (Fig. 4). The horseshoe is broadly rounded ventrally and has shallow lateral indentations near the top. The sharply-pointed ears are slightly longer than the head with a large, though shallow, antitragus. The wings arise from the ankles, and the interfemoral membrane from half-way between the knee and the ankle.

**Habits**: Colonial, it roosts in dense clusters on walls and ceilings of caves or in narrow cracks. It hibernates, except possibly in warm environments.

**Habitat**: Wooded foothills and mountains, in damp caves with high ceilings.

**Distribution**: Southern Europe, Asia Minor, Iran and North Africa. The subspecies *Rhinolophus euryale judaicus* occurs in Lebanon, Jordan and in Israel, where it is known from the Sea of Galilee, the coastal plain and the Judean hills.

## Mehely's Horseshoe Bat

Plate 6, Figure 44

*Rhinolophus mehelyi* Matschie, 1901
Parsav Hiver (Hebrew)
Khaffash (Arabic)

**Description**: Head and body 409–60.4 mm.; ear 18–23 mm.; forearm 43–52.9 mm.; wingspread 229–340 mm.; tail 20.3–20.9 mm.

Mehely's Horseshoe Bat is medium-sized, pale grayish-buff above and buffy-white below. It has dark gray-brown "spectacles" around the eyes. It is distinguished from *Rhinolophus euryale* by its nose-leaf, in which the lancet tapers abruptly about half-way between the tip and the dorsal connecting point of the sella. The distal half of the lancet is narrow and parallel-sided in the anterior view. The upper connecting process of the sella is blunt, not pointed, and much longer than the lower connecting process (Fig. 3). The horseshoe is narrowly rounded with a rather deep lateral indentation near the top. The second phalanx of the fourth finger is more than twice the length of the first phalanx.

**Habits**: Unknown, except that it emerges at dusk. It may associate with other species of horseshoe bats and Schreiber's Bat. Its flight is slow with short gliding and low over the ground.

**Habitat**: Caves, in areas close to water.

**Distribution**: Transcaucasia, Iran, Turkey and Lebanon. In Israel, it has been recorded from the north (Rosh Hanikra), the coastal plain (Herzlia), Galilee and the Judean hills (Jerusalem).

# LEAF-NOSED BATS: HIPPOSIDERIDAE

The Leaf-nosed Bat lacks a defined horseshoe. The lancet is a transverse leaf with three parts and no sella. The ears are large, pointed and lack a tragus. Leaf-nosed bats inhabit the Old World tropics and are represented in Israel by a single species.

## Trident Leaf-nosed Bat
Plate 6

*Asellia tridens tridens* E. Geoffroy, 1813
Parsafon (Hebrew)
Khaffash (Arabic)

**Description**: Head and body 37–65 mm.; ear 14–20.2 mm.; forearm 43.5–54.6 mm.; wingspread about 280 mm.; tail 16–28.5 mm.

The Trident Leaf-nosed Bat is of medium size, with two color phases, a pale grayish-brown or light orange-brown above, with whitish underparts. It is distinguished from other leaf-nosed bats by the nose-leaf, which has no median process hiding the nostrils, and the upper hind portion does not terminate in a single point but is divided into three distinct points, the central one being the highest. The ears are large and separated, but the antitragus is small. The tail extends a short way beyond the interfemoral membrane.

**Habits**: It appears late after dusk. Its flight is low, swift and butterfly-like. It tolerates high temperatures, at least 38° C.

**Habitat**: Caves, abandoned buildings and tombs.

**Distribution**: North Africa, Senegal, Egypt to Zanzibar, southern Iran, southern Syria, Arabia, eastward to northwestern India. In Israel, it is common and widespread from Mount Carmel, the coastal plain, Jordan Valley, Judean hills to the Arava Valley (north of Eilat).

# FREE-TAILED OR SHARP-NOSED BATS: MOLOSSIDAE

The Free-tailed, Sharp-nosed or Bull-dog Bat is characterized by the distal half of the thick tail which extends beyond the hind edge of the interfemoral membrane. The muzzle is broad, truncated, and the upper lip is grooved with vertical wrinkles. The ears are large, broad and thick, with a distinct tragus which is more or less united. The legs are short and well-developed with long, curved hairs. Free-tailed bats are found in the warmer and tropical parts of both hemispheres. In Israel, they are represented by a single species.

## European Free-tailed Bat      Plate 4

*Tadarida teniotis rueppelli* Temminck, 1826

Ashaf Matsui (Hebrew)

Khaffash (Arabic)

**Description:** Head and body 123–133 mm.; ear 25–32 mm.; forearm 58–63.9 mm.; wingspread about 387 mm.; tail 46.6–54.8 mm.

The Free-tailed Bat is distinguished by its fairly large size; short, velvety fur; and sooty-brown color, with slightly paler underparts. Juveniles are grayer. The muzzle is broad and truncated, with thick, wrinkled lips, and without a nose-leaf. It has large, broad ears with a short tragus and project to the side. The tail, which penetrates the interfemoral membrane, projects freely above it, well beyond the hind margin. The wings are very long and narrow. At rest, the third joint of the first finger folds backward instead of forward as in other bats. Between a third and a half of the tail protrudes beyond the short membrane. The calcar is without a post-calcareal lobe.

**Habits:** More terrestrial than other bats, it runs about almost like a mouse. Its flight is long, fast and direct without fluttering. Unlike most other bats, the young is carried on the back of the mother.

**Habitat:** Mountains with gorges and cliffs up to 1,920 m. (Alps) and near human settlement. It roosts in caves and crevices in warm areas.

**Distribution:** Southern Europe, Madeira, the Canary Islands, through North Africa to southern China, Taiwan and Japan. The subspecies *T. teniotis rueppelli* is known from Egypt, Lebanon and Iraq. In Israel, it is found on Mount Carmel, the coastal plain, Jordan Valley, Judean hills and the Negev Desert, as far south as Eilat.

# PLAIN-NOSED BATS: VESPERTILIONIDAE

The plain-nosed bats are characterized by their simple snout, the muzzle usually having no appendages. The ears vary greatly in size, but there is always a tragus present, usually long and pointed. All have a large interfemoral membrane which either completely encloses the tail, or without the last vertebra projecting. Several species are difficult to identify, particularly the Myotis group, even from close up. They may live remarkably long, more than thirty years. Plain-nosed bats are the largest and most widespread family of bats. They are nocturnal or crepuscular and found in temperate and tropical regions of the world. In Israel, they are represented by eleven species.

## Common Noctule or Great Bat                Plate 8

*Nyctalus noctula lebanoticus* Harrison, 1962

Rmashan Leyli (Hebrew)

Khaffash (Arabic)

**Description:** Head and body 50–100 mm.; ear 14–19 mm.; forearm 40–70 mm.; wingspread 343–368 mm.; tail 35–65 mm.

The Common Noctule is medium-sized though robust, with a broad muzzle, dark golden-brown above and paler below. In Great Britain, pied and almost black variations occur. The juvenile is much darker than the adult. It is distinguished by its widely separated, short, dark brown ears, which are broader than long; low antitragus; and club-shaped tragus. The forearm is massive, and the wing long and slender. The calcar is well-developed, longer than the rather short tibia. The last joint of the tail is free.

**Habits:** In Europe, it is active throughout the year, except late December and January, before sunset or at twilight and again at dawn, and sometimes also during the day. A single young is born towards the end of June. It hibernates from October to the end of March. Its flight is high, quick and dashing, often gliding down obliquely on expanded wings. Its voice is shrill, clear and cricket-like.

**Habitat:** Woodlands, desert, tree holes, fissures and crevices in caverns, mainly in lowlands, but also on mountains up to 1,600 m.

**Distribution:** Widespread in western Europe, Russia, south to Asia Minor, eastwards through Iran and Asia. The subspecies *N. n. lebanoticus* is rare in the region where it is known from Lebanon and Israel (near Jericho).

## Greater Mouse-eared Bat                Plate 3

*Myotis myotis macrocephalicus* Harrison and Lewis, 1961

Nishpon Gadol (Hebrew)

Khaffash (Arabic)

**Description:** Head and body 65–84 mm.; ear

27–29 mm.; forearm 58.1–71 mm.; wingspread about 3,005 mm.; tail 52–63 mm.

The Large Mouse-eared Bat is distinguished by its fairly large size and color — light ochrous-brown above and paler below. It has large, broad ears, with 7–8 transverse folds, the front edge distinctly bent backwards. A long, pointed tragus is less than half the height of the ear. The wings are broad; the membrane reaches the base of the foot; and the tip of the tail is free.

**Habits:** It appears late and flies for 4–5 hours at a height of 5–8 m. Its flight is slow and in a straight line, direct with "rowing" wing beats. When alighting, it does a somersault. Females are colonial, whereas males are solitary. It sleeps far from the entrance to its quarters, usually hanging freely. It occasionally migrates.

**Habitat:** Caves, grottoes, hollow trees, dome roofs, rarely in crevices; in open woodland and cultivated lands. It winters in caves and summers in attics.

**Distribution:** Western Europe, south Asia Minor, Russia, east to Carpathians, Arabia. The subspecies *M.m. macrocephalicus* is found in Syria and Lebanon. In Israel, it is fairly common in Upper Galilee.

## Lesser Mouse-eared Bat          Plate 3

*Myotis blythi omari* Thomas, 1906
Nishpon Matsui (Hebrew)
Khaffash (Arabic)

**Description:** Head and body 115–139 mm.; ear 21.4–28 mm.; forearm 57–63.8 mm.; wingspread under 305 mm.; tail 51–65 mm.

The Lesser Mouse-eared Bat is distinguished by its medium size; pale tawny upper parts; and whitish underparts. It has a smooth, pointy snout and smallish ears, the outer edge with 5–6 transverse folds and the front edge moderately bent backwards, with a long, pointed tragus. The hind legs and tail are comparatively long. The interfemoral membrane reaches the end of the tail.

**Habits:** Emerging late in dusk or after dark, its flight is slow and even. It often associates with Schreiber's and horseshoe bats. Sometimes it migrates.

**Habitat:** Caves, tombs, hollow trees and empty buildings.

**Distribution:** Mediterranean region of Europe, southern Russia, eastwards through southwest Asia, Asia Minor, Iran, Turkey, Syria, Arabia and North Africa. The subspecies *M. b. omari* is known from Iran, Turkey, Iraq, Syria and Lebanon. In Israel, it has been recorded from the Jezreel valley (Beit She'arim).

## Geoffroy's or Notch-eared Bat          Plate 7

*Myotis emarginatus emarginatus* Geoffroy, 1806
Nishpon Pgoom Ozen (Hebrew)
Khaffash (Arabic)

**Description:** Head and body 75.2–95 mm.; ear 13.8–16.8 mm.; forearm 36.7–42.9 mm.; wingspread 229–267 mm.; tail 35–44 mm.

Geoffroy's Bat is of medium size, with rufous upper parts and pale ochre underparts. It is distinguished by its small ears; 6–7 transverse folds; with a fine notch on the outer edge, about half the height of the ear and a long, pointed tragus. The juvenile is gray, and the ears are less emarginated. The wing membrane extends to the base of the toes. The short calcar reaches only about half the distance between the ankle and tail along the margin of the interfemoral membrane. The extreme tip of the tail protrudes beyond the membrane.

**Habits**: It emerges late, flying 1–5 m. above the ground or over water. It is mainly sedentary, often associating with horseshoe bats.

**Habitat**: Caves and tombs in woodlands, near water.

**Distribution**: Southern and Central Europe, from France to Italy, Lebanon and eastern Iran. In Israel, it is common on Mount Carmel and in the coastal plain.

## Long-fingered Bat                Plate 9

*Myotis capaccinii bureschi* Heinrich, 1936
Nishpon Gadot (Hebrew)
Khaffash (Arabic)

**Description**: Head and body 44–51.3 mm.; ear 11.8–14.6 mm.; forearm 39.8–41.8 mm.; wingspread about 229 mm.; tail 35–46 mm.

The Long-fingered Bat is distinguished by its small size; small ears with a pointed tragus of medium length; and relatively large foot, three-quarters of the length of the tibia. The

upper parts are light brown, while the belly is whitish. The calcar is straight and about a third of the length of the tail membrane, on which there are hairs.

**Habits**: It emerges at late dusk. Its wingbeats are shallow, the downstrokes not below the level of the body. It sometimes associates with mouse-eared bats. It catches insects, especially over water, by using its tail membrane and/or feet.

**Habitat**: Caves and crevices, in wooded areas, often near water.

**Distribution**: Mediterranean region, Spain, southern France, Italy, Corsica, Switzerland, Sardinia, Transylvania, Yugoslavia, Greece, Bulgaria, Russian Turkestan, Iran, Arabia, Cyprus, Asia Minor, Morocco and Algeria. The subspecies *M. c. bureschi* is known from Bulgaria. In Israel, it has been recorded from around the Sea of Galilee and Mount Carmel.

## Natterer's Bat                Plate 9

*Myotis nattereri hoveli* Harrison, 1964
Nishpon Dak Ozen (Hebrew)
Khaffash (Arabic)

**Description**: Head and body 79.5–94 mm.; ear 14–18.3 mm.; forearm 38.2–40.3 mm.; wingspread reaches 280 mm.; tail 36.7–43.6 mm.

This smaller, paler race of Natterer's Bat is medium-sized, with light grayish-brown upper parts and dull, whitish underparts, but there is much variation, and paler specimens are not uncommon. The young are darker than the adult. It is characterized by its plain, elongated muzzle, very tall ears and long, attenuated tragus, more than half the length of the ear. The wing membrane extends to the base of the toes. The middle and free hind margin of the membrane forms an obtuse angle between the legs and is distinctly fringed with stiff hairs along the lower edge. The tip of the tail either projects beyond the margin of the interfemoral membrane or is completely withdrawn. The calcar is short and reaches half the distance between the ankle and the tail.

**Habits**: It emerges for flight from several hours before sunset to about twenty minutes after. The flight is short, lasting only about twenty minutes before resting, but continuing throughout the night until a few minutes before sunrise. It usually flies low and deliberately brushes against leaves in flight to disturb insects. The flight is slow and steady with the tail held straight out. The wings appear light compared to the body. It alights with its head either up or down. It is generally sedentary.

**Habitat**: Open woodlands, farmland, near open water or marsh, in cracks and caves with high humidity.

**Distribution**: Europe, the Middle East and Far East. The subspecies *M. n. hoveli* is known only from Israel, where it is common in Galilee, the coastal plain and the Judean hills.

## Serotine Bat Plate 7

*Eptesicus serotinus serotinus* Schreber, 1774
Aflul Matsui (Hebrew)
Khaffash (Arabic)

**Description**: Head and body 119–133.5 mm.; ear 18.2–21 mm.; forearm 53.2–56 mm.; wingspread 305–356 mm.; tail 47–58 mm.

The Serotine Bat is distinguished by its fairly large size, with dark, rich brown upper parts and paler underparts. Its face is bare except for a few glandular hairs, and it has broad, rounded ears with five transverse folds and a short, blunt tragus, less than half the length of the ear. The wings are broad, with a very small calcareal lobe. The tip of the tail projects slightly but noticeably beyond the interfemoral membrane.

**Habits**: It appears soon after sunset and again at dawn. Its flight is slow and fluttering, with a wide beat. It usually flies about 9–12 m. above the ground, often swooping down in a sudden oblique dive. Males are usually solitary, but sometimes mix with females in small groups. It hibernates, but may be active on warm nights.

**Habitat**: Open areas sheltered by trees. Caves, hollow trees, cracks in walls or under roofs.

**Distribution**: This species has the widest range of all bats — being the only bat common to both the Old and New Worlds. It is widespread in central and southern Europe, east toward Asia, Lebanon, North and part of West Africa. In Israel, it is found in Upper Galilee, the coastal plain and the Judean hills.

## Botta's Serotine Bat, Lesser Serotine Bat

Plate 7

*Eptesicus bottae innesi* Lataste, 1887
Aflul HaNegev (Hebrew)
Khaffash (Arabic)

**Description**: Head and body 99.5 mm; ear 16.5 mm.; forearm 42.3 mm.; tail 40 mm.

The Lesser Serotine Bat is of medium size. The upper parts are a pale buffy-gray, and the underparts are white tinged with buff, and contrast strongly with the dark, blackish ears, wings and tail membrane. The ears are rounded with a short, blunt tragus, less than half the height of the ear. A small but distinct post-calcareal lobe is present. The tail is relatively long.

**Habits**: Its flight strong, noisy and moderately high. It is attracted by insects around electric lights.

**Habitat**: Desert oases with trees and cultivated fields.

**Distribution**: Egypt. In Israel, it is very rare but has been recorded in the Arava Valley (from Ein Gedi and Yotvata).

## Common Pipistrelle

Plate 9

*Pipistrellus pipistellus pipistrellus* Schreber, 1774
Atalfon Aeropi (Hebrew)
Khaffash (Arabic)

**Description**: Head and body 70–76 mm.; ear 10–12 mm.; forearm 30–31 mm.; wingspread more than 203 mm.; tail 30–35 mm.

The Common Pipistrelle is small, though robust, with dark brown upper parts and a lighter brown belly. The ears are short and broad, with a minute lobular antitragus. The tragus is almost half the height of the ear and its distinctive feature is a deep basal notch on the posterior border. The calcar is more than half the border of the interfemoral membrane, and the post-calcareal lobe is large. The tail is shorter than the head and body, and its extreme tip projects from the membrane.

**Habits**: It emerges soon after sunset and is very active throughout the night, for a longer period than any other species, from March until winter. Even in winter, if the weather is mild, it may emerge for an hour. It usually flies fast and jerkily, 2–10 m. above the ground. It often feeds near water, in gardens and under streetlights. It will eat as many as 3,000 insects in one night. Males are solitary or in small groups.

**Habitat**: Human habitation, such as any crevices behind water pipes and gutters, between woodwork and brickwork; edges of woods, agricultural land and rarely in caves.

**Distribution**: The subspecies *P. p. pipistrellus* is known from Europe and parts of Asia, Asia Minor, Lebanon and Morocco. In Israel, it is very rare in the northern part of the country on Mount Hermon and the Golan Heights and seems to reach the southern limit of its range in Upper Galilee.

## Kuhl's Pipistrelle — Plate 9

*Pipistrellus kuhli ikhwanius* Cheesman and Hinton, 1924

Atalfon Leven Shulayim (Hebrew)

Khaffash (Arabic)

**Description**: Head and body 65–91 mm.; ear 9.5–14.5 mm.; forearm 29.6–36.2 mm.; wingspread about 204–229 mm.; tail 29.3–42 mm.

Kuhl's Pipistrelle is small- to medium-sized, plain-nosed, with variable russet or mouse-brown upper parts and ashy-white underparts. The juvenile is darker than the adult. The brownish-black ears are shorter than the head, with a pronounced tragus, rounded but not enlarged at the tip, about half the height of the ear and a distinct notch at the base of the posterior border. The relatively narrow wings are distinctive for the slender, white margin along the trailing edge of the wing which invades the membrane. The calcar is long with a distinct semi-oval lobe outside its base.

**Habits**: It appears before sundown and feeds around street lights and over water. Its flight is not very high and erratic, with abrupt turns and sudden dives.

**Habitat**: Oak forests and around settlements, where it seeks narrow spaces or cracks in buildings; cliffs, tombs and date palms.

**Distribution**: Widespread from southern Europe, southern Russia, southwest Asia, Asia Minor, Arabia, northern Sinai and North, East and South Africa. The subspecies *P. k. ikhwanius* is known from Lebanon, Syria, Iraq, Arabia and Egypt (Sinai). In Israel, it is the commonest insectivorous bat and found throughout the north into the northern Negev Desert, Judean Desert and around the Dead Sea.

## Rueppel's Pipistrelle — Plate 9

*Pipistrellus rueppellii coxi* Thomas, 1919

Atalfon Rippel (Hebrew)

Khaffash (Arabic)

**Description**: Head and body 46 mm.; ear 11 mm.; forearm 30.3–32.2 mm.; tail 33 mm.

Rueppel's Pipistrelle is rather small, sandy-buff above, which contrasts strongly with its blackish ears, limbs and tail, and pure white below, sharply divided on the sides of the neck. The ears are tall and rather narrow, with a distinct notch at the base of the external border, separated from a small, low antitragus. The tragus is half the height of the ear, tall and curved, with a blunt tip and distinct, triangular basal lobe. The wing membrane originates high on the metatarsal. It has a remarkably long penis — 13 mm.

**Habits**: It is nocturnal. Otherwise, it is unknown.

**Habitat**: Near marshes, human habitation, palm groves and gardens.

**Distribution**: Egypt, Sudan, Kenya, Uganda, Tanzania, the Congo, northern and southern Zimbabwe, Nyasaland, Ngamiland, Bechuanaland, Senegal and Angola. The subspecies *P. r. coxi* is known from Iraq. In

Israel, it has been recorded from northernmost Upper Galilee.

## Savi's Pipistrelle                              Plate 8

*Pipistrellus savii caucasicus* Satunin, 1901
Atalfon Savi (Hebrew)
Khaffash (Arabic)

**Description**: Head and body 37–55 mm.; ear 10–11 mm.; forearm 31.8–36.5 mm.; tail 30–32 mm.

Savi's Pipistrelle is small- to medium-sized, with variable, pale buffy-brown upper parts and a buffy-white belly. The ears are rather large and black. The base of the external border is separated from a low antitragus by a shallow notch. The tragus is less than half the height of the ear and very wide, slightly curved forward, with a bluntly triangular, well-developed basal lobule. The outer edge has four transverse folds. The base of the outer edge of the ear has two serrations, one above the other. In the dentition, the first upper incisor is bicuspid, and the second upper incisor is very tall, three-quarters of the height of the first incisor and almost reaching the height of its secondary cusp. The wing membrane originates at the base of the outer toe. There is a narrow, post-calcareal lobe, with 3–5 mm. of the calcar tip without membrane.

**Habits**: It appears soon after sunset and is active all night. Its flight is straight and not very fast. It is probably migratory and hibernates alone.

**Habitat**: The edge of mountain meadows and wood up to 2,600 m. above sea level, desert oases in sandy deserts with acacia trees and cultivated areas bordered with trees. It roosts in rock crevices, tree hollows, caves and buildings.

**Distribution**: Southern Europe and Russia, Asia Minor and Lebanon east to Siberia and Mongolia, North Africa, but not Egypt. The subspecies *P. s. caucasicus* is known from Lebanon. In Israel, it has been recorded from Upper Galilee.

## Bodenheimer's Pipistrelle                    Plate 9

*Pipistrellus savii bodenheimeri* Harrison, 1960 n. status
Atalfon Bodenheimer (Hebrew)
Khaffash (Arabic)

**Description**: Head and body 31.7–40 mm.; ear 8.4–11.6 mm.; forearm 28–32.8 mm.; tail 26–37 mm.

Bodenheimer's Pipistrelle is very small and delicate, with very pale, whitish-buff upper parts and a whitish belly. The ears are relatively large, pale brownish and translucent. The well-developed antitragus is narrow and high, sharply separated from the external corner of the ear. The tragus is about half the height of the ear, with the posterior border sharply angulated, coming to a blunt point, and concave above a small basal lobule. In the dentition, the first upper incisor is bicuspid or has an incipient second cusp. The second upper incisor is very tall, three-quarters of the height of the first incisor, and higher than its secondary cusp. The calcar is weak, supporting less than half of the border of the membrane; the post-calcareal lobe is small but distinct. The dusky-black wing membrane originates at the base of the outer toe and contrasts with the pale back. The tip of the tail protudes from the interfemoral membranes.

**Habits**: Active at dusk, its flight is very delicate, close to the ground and around trees. It roosts in crevices. In April, it was found pregnant with two embryos. It hibernates between October and April.

**Habitat**: On mountains to below sea level, near desert oases, cultivated land bordered by trees and surrounded by sandy desert.

**Distribution**: Egypt (Sinai), Arabian peninsula, south to Aden. In Israel, it is common near the western shore of the Dead Sea and in the southern Arava Valley.

**Taxonomic Remarks**: *Pipistrellus bodenheimeri* has been described as a small and very distinctive Arabian representative of the *P. savii* group, as indicated by its dentition (Harrison, 1960). Its dentition is nearly the same as that of *P. savii*, which is variable; its small size and pale coloration are adaptations to the desert; there is no significant morphological gap between *P. bodenheimeri* and *P. savii* and it is nowhere sympatric. It therefore appears that *P. bodenheimeri* is a geographical and ecological subspecies of *P. savii*. It is suggested, therefore, that it be known as

*Pipistrellus savii bodenheimeri* n. status.

## Desert Pipistrelle, Pygmy Pipistrelle Plate 8

*Pipistrellus ariel* Thomas, 1904

Atalfon Ariel (Hebrew)

Khaffash (Arabic)

**Description**: Head and body 28.4–31.7 mm.; ear 9.2 mm.; forearm 28–31 mm.; tail 33mm.

The Desert Pipistrelle is distinguished by its small size; grayish-brown upper parts; narrow, pointed muzzle; and tall, narrow ears with a well-developed tragus and poorly-developed antitragus. In its dentition, the first upper incisor is unicuspid, and the second upper incisor is long, almost two-thirds the length of the first incisor. The tip of the tail protrudes 1.3 mm. from the interfemoral membrane.

**Habits**: It is nocturnal. Otherwise, it is unknown.

**Habitat**: Desert.

**Distribution**: Eastern Egypt and Sudan. It is very rare, but has been recorded in Israel southwest of the Dead Sea (Nahal Ze'elim) and in the Arava Valley.

**Taxonomic Remarks**: According to Yom-Tov et al., (1992), *P. ariel* is probably conspecific with *P. bodenheimeri*. The first upper incisor, which is diagnostic in pipistrelles, is unicuspid in *P. ariel*, and not bicuspid, as in *P. savii bodenheimeri* comb. nov.

## Hemprich's Long-eared Bat — Plate 5

*Otonycteris hemprichi jin* Cheesman and Hinton, 1924

Udnan (Hebrew)

Khaffash (Arabic)

**Description**: Head and body 63–130 mm.; ear 31.5–42 mm.; forearm 61–66 mm.; tail 50.5–58 mm.

Hemprich's Long-eared Bat is large, very pale grayish buff above with a white belly, distinctly demarcated at the sides of the neck. The very large, pale, translucent ears have a small antitragus, sharply separated from the basal part of the outer border. The tragus is very large, about half the height of the ear. The wing membranes are pallid and translucent, and they originate at the base of the outer toe. The calcar is weak, supporting less than half the border of the membrane with no post-calcareal lobe. The tail is shorter than the head and body, and its tip projects 4–5 mm.

**Habits**: Its flight slow and flopping. It inhabits crevices behind rocks on steep sides of wadis. It emits a long buzzing sound at rest.

**Habitat**: Rocky deserts with acacia trees.

**Distribution**: Kashmir, Russian Turkestan through Iran and Iraq. In Israel, the subspecies *O. h. jin* is known from the western shore of the Dead Sea, the Negev Desert and the Arava Valley.

## Arabian Barbastelle — Plate 8

*Barbastella barbastella leucomelas* Cretzschmar, 1826 n status

Blumaf Shahor (Hebrew)

Khaffash (Arabic)

**Description**: Head and body 43–60 mm., females, larger than males; ear 6–7.5 mm.; forearm 35–45 mm.; wingspread about 265 mm.; tail 40–55 mm.

The Arabian Barbastelle is essentially similar to the Eurasian race, although paler. It is small, dark brown or blackish, with a whitish or yellowish "frosted" appearance on the back and a paler, buffy-white abdomen. Juveniles are darker without the whitish tips. It has peculiar, moderately large ears, which are nearly square, facing forwards, with the anterior margins close together and joined on the forehead. The blackish ears lack the small, projecting lobule in the middle part, usually found in the Eurasian race. There is no antitragus. The tragus is more than the height of the ear.

**Habits**: Little is known about the Arabian subspecies. In Europe, the sexes segregate. It emerges early in the evening, even before sunset. Its flight is slow and flapping, near the ground or fairly high. It occasionaly migrates and hibernates.

**Habitat**: Caves in drier regions where it hibernates from September to early April; crevices of tree trunks, walls, in roofs and behind shutters of houses.

**Distribution**: The nominate subspecies is

found in Europe and Russia. The Arabian race *B. b. leucomelas* occurs in the Caucasus, Transcaucasia, Russian Turkestan, Chinese Turkestan, southern Asia, Iran and Egypt (Sinai). In Israel, it has been recorded from the southern Arava Valley (Eilat), but is very rare

**Taxonomic Remarks**: The small projecting lobule in the middle part of the ear that distingishes *B. barbastella* from *B. leucomelas* is sometimes lacking in *B. barbastella*, and therefore not diagnostic. It is suggested, therefore, that *B. leucomelas* is a southern subspecies of *B. barbastella* and that it be known as

*B. barbastella leucomelas* n. status.

## Gray Long-eared Bat     Plate 5

*Plecotus austriacus christiei* Fischer, 1829
Aznan (Hebrew)
Khaffash (Arabic)

**Description**: Head and body 39.8–49 mm.; ear 31.5–39 mm.; forearm 37.8–41 mm.; tail 41.3–50 mm.

The Gray Long-eared Bat is medium-sized with very large ears, pale brown upper parts and chest and a white belly. The young are darker. It is characterized by its enormous, pale brown, semi-translucent ears, which are nearly as long as the head and body combined and joined across the forehead. The antitragal lobe is virtually absent. The tragus is about half as high as the ear and 5.5–6 mm. at its widest point. The ears can fold, leaving only the tragus sticking out. This may prevent damage and

reduce heat loss. The wings are fairly short and broad. The thumb with the claw is usually more than 6 mm. The calcar is large and supports about half the border of the membrane. There is no post-calcareal lobe. The tail is relatively long, equal to or exceeding the length of the head and body.

**Habits**: At rest, the ears are in constant motion. Appearing about a half-hour after sunset, it seems to hunt throughout the night. Its flight is rapid and irregular, with its ears held horizontally forward. At times, it glides and hovers, often picking insects off leaves and blossoms. It may be recognized by its high-pitched notes. In winter, it hibernates.

**Habitat**: Caves in barren mountain valleys, hollow trees and houses.

**Distribution**: Europe, Asia Minor, Iran, Afghanistan and Kashmir, North and East Africa, Egypt. The subspecies *P. a. christiei* apparently occurs in Egypt, Sinai and Syria. In Israel, it is widespread in the north, the Judean hills (near Jerusalem) south through the Negev Desert to Eilat.

## Schreiber's Bat, Long-winged Bat,     Plate 9 Bent-winged Bat

*Miniopterus schreibersi schreibersi* Kuhl, 1819
Kanfan (Hebrew)
Khaffash (Arabic)

**Description**: Head and body 58–61.5 mm.; ear 8.3–12 mm.; forearm 44–46.7 mm.; tail 52–61.2 mm.; wingspread 344 mm.

Schreiber's Bat is fairly large, and distinguished by its variable grayish or pale brown color above; grayish-white below; high crown; low, rounded ears; and long, narrow wings. The short, rounded tragus is less than half the height of the ear, and the antitragus is low and not well-defined. The second joint of the third finger is very long, almost three times the length of the first joint. At rest, the fingers bend differently from other bats. The feet and tail are relatively long. The calcar is about a third to a half the length of the interfemoral membrane, and there is no post-calcareal lobe.

**Habits**: Appearing soon after sunset, it continues to fly late into the night. Its flight is very high, rapid, sustained and swallow-like, 5–20 m. above ground, and is accompanied by much squeaking. It migrates, but not very far, and hibernates in caves.

**Habitat**: Slopes of high mountains and caves.

**Distribution**: The species is widespread in the Old World, through Europe and Asia, Asia Minor, Iran, Iraq and Lebanon. In Israel, it has been recorded from the central coastal plain, south to Lahav.

# CARNIVORES: CARNIVORA

The order of carnivores is characterized by the large, blade-like canine teeth and usually also by strong claws. Carnivores have a keen sense of sight, smell and hearing and are generally predators, well-adapted for catching and killing live prey. Some are omnivorous, such as bears and to some extent badgers; whereas hyaenas feed largely on carrion. Carnivores are widespread throughout the world and are represented in Israel by seven families and three subfamilies.

Fig. 7. Marbled Polecat in defensive position.

# DOGS, WOLVES, JACKALS AND FOXES: CANIDAE

The family of canines includes domestic dogs, wolves, jackals and foxes. They are all dog-like in appearance, and are characterized by strong, muscular bodies; elongated muzzles; four toes on the hind feet; and nonretractile claws. Their limbs vary in length, but are rarely short in relation to the body. Their hairy or bushy tails are usually of moderate length. Foxes have vertical pupils in bright light. They are smaller, more slender, with shorter legs; a long, bushy tail; and relatively larger ears. The Domestic Dog, *Canis lupus familiaris*, was bred from the wolf.

Wolves and foxes are cosmopolitan and inhabit a great range of environments, from the Arctic to the tropics and desert, whereas jackals are confined to the Old World. In Israel, there is one species of wolf, one species of jackal and three species of foxes. Qumsiyeh (1996) claims four species of foxes occur in the Holy Land. He includes the Fennec, *Vulpes zerda*, for which there is only speculation, but no reliable records.

## Indian Wolf                    Plates 10, 11

*Canis lupus pallipes* Sykes, 1831

Ze'ev (Hebrew)

Theeb (Arabic)

**Description**: Head and body 760–1,200 mm.; ear 105–125 mm.; hind foot 220–255 mm.; tail 390–430 mm.

The Indian Wolf is a medium-sized race of the larger Eurasian Wolf. Among wolves, it is closest to the dog. It is distinguished by its fairly large size; short fur, grayish-brown upper parts, sometimes with blackish on its back. The muzzle, cheek and throat are white. There is ruff-like hair around the cheeks and the neck. Darker specimens have a distinct white spot on the cheek. The backs of the ears are a light rufous. A blackish streak on the outer tarsus is variably pronounced. The relatively short, rope-like tail reaches the hock and has a black tip. There is considerable variation in color. On average, it is larger and darker in the north and smaller and paler in the south.

**Habits**: Mostly nocturnal, it associates in packs of a dozen or less. It forages over a large territory. When running, it holds its tail in line with its body. An opportunistic predator, it preys on wildlife and occasionally on livestock. Also a scavenger, it frequents garbage dumps. It interbreeds with feral dogs. After a gestation period of about 63 days, 3–13 cubs are born in a den. Usually silent, it rarely howls, but sometimes barks.

**Habitat**: Open rocky hills, plains and desert steppes, but not dense forests.

**Distribution**: Widespread from northern India

to Sind, south to Dharwat, Baluchistan, southern Iraq, Kuwait, northern Arabia, Lebanon and Syria. In Israel, the subspecies *C. l. pallipes* is extirpated from the coastal plain, but still occurs in the Judean hills, and is an intruder in the Huleh Valley from the Golan Heights. A slightly smaller and paler population appears to inhabit the northern Negev Desert and northern Arava Valley.

## Arabian Wolf                         Plate 10, Frontispiece

*Canis lupus arabs* Pocock, 1934

Ze'ev Aravi (Hebrew)

Theeb (Arabic)

**Description**: Head and body 700–800 mm.; ear 80–92 mm.; hind foot 184–197 mm.; tail 300–340 mm.

A desert race of the Eurasian Wolf, the Arabian Wolf is similar to the Indian race but smaller, leaner, and usually paler with short, thin fur, particularly in summer. It resembles an Alsatian or police dog. In Israel, the pads of the third and fourth toes are often connected in the back. This helps distinguish their tracks from those of dogs.

**Habits**: It is active at sunset and at night. It is less gregarious than the Indian Wolf and associates in small numbers — 2–7. It feeds on jirds, hares, partridges, gazelles and rarely livestock and is a scavenger at garbage dumps. Otherwise, it is similar to the Indian Wolf.

**Habitat**: Desert mountains, wadis and plains in open areas, not far from water.

**Distribution**: Saudi Arabia, Oman, Kuwait (where it may intergrade with the Indian subspecies), and Egypt (the southern and eastern Sinai desert. The smaller *C. lupus. lupaster* of Egypt reaches the northwestern Sinai). In Israel, *C. lupus. arabs* inhabits the southern Arava Valley and appears to intergrade with the Indian subspecies in the northern Negev Desert and northern Arava Valley.

**Taxonomic Remarks**: Qumsiyeh (1996) claims that morphological evidence is not yet available to determine that the Great Egyptian Jackal is indeed a wolf. He fails to note that there are significant morphological characters which align *C. l. lupaster* with *C. lupus* and not with *C. aureus*. Measurements of the skull length, mandible and carnassial of *C. l. lupaster* overlap the lower limits of *C. lupus arabs*, but show a distinct mensural gap with those of *C. aureus*. The so-called Great Egyptian Jackal, recorded from Palestine by Flower (1932) and Bate (in Bodenheimer, 1958), is in fact a small wolf, *C. lupus lupaster*, and does not occur in Israel (Ferguson, 1981a).

## Syrian or Asiatic Jackal,                         Plates 11, 12
## Golden Jackal

*Canis aureus syriacus* Hemprich and Ehrenberg, 1833

Tahn (Hebrew)

Wa Wie (Arabic)

**Description**: Head and body 600–900 mm.,

female smaller than male; ear 73–89 mm.; hind foot 140–162 mm.; tail 210–270 mm.

The Syrian Jackal is a small race of the Asiatic Jackal. It is smaller than a wolf, with relatively shorter legs and tail. It is larger than a fox and can be distinguished by its relatively smaller, rufous ears and shorter, black-tipped tail. The upper parts are usually yellowish-gray tinged with rufous, grayer on the back, which is grizzled with varying amounts of black. There is also a reddish phase (Plate 12). The underparts are almost white. There are two dark bands across the lower throat and upper breast. The winter coat is longer and grayer. Its voice is a long cry or howl which ends in a series of yelping notes, usually emitted just after dark or at dawn.

**Habits**: The jackal walks with a slouching gait. It travels singly or in pairs, sometimes in small family packs. The gestation period varies around 60–63 days, when 3–4 pups are born in a chamber at the end of a burrow or in some natural grotto. Mostly a scavenger, it may attack small wildlife, injured or young gazelles, lambs and poultry.

**Habitat**: Hills, plains, around orange groves, in forests and on the outskirts of towns and villages.

**Distribution**: Southern Europe, North Africa, Egypt, Asia Minor, Arabia, to India and the Indochinese peninsula. The subspecies *C. a. syriacus* is common throughout the northern half of Israel to just south of Beersheba, but does not penetrate the desert.

# Arabian Jackal                                    Plate 12

*Canis aureus hadramauticus* n. status
Tahn Aravi (Hebrew)
Wa Wie (Arabic)

**Description**: Head and body 710–753 mm.; ear 70–89 mm.; hind foot 147–160 mm.; tail 220–270 mm.

The little known Arabian Jackal is smaller than the Syrian subspecies, with less black on the back and reduced red on the backs of the ears and legs. The os penis is distinctly smaller and shorter than that of the Syrian subspecies.

**Habits**: Presumably similar to those of the Syrian race.

**Habitat**: Arid regions near oases in humid enclaves with reed beds and bushes near springs.

**Distribution**: Southern Arabia. In Israel, jackals found near the Dead Sea (Ein Feshcha and Neot Hakikar) probably belong to this subspecies.

**Taxonomic Remarks**: *Canis hadramauticus* Noack, 1896, was first described as a distinct species. It was provisionally assigned to *C. a. aureus* by Harrison (1968). Since the Arabian Jackal is indistinguishable from the species *C. aureus*, but morphologically and geographically distinct from *C. a. syriacus*, it is suggested that it be given a subspecific rank and be known as

*Canis aureus hadramauticus* n. status

## Common Red Fox — Plates 11, 13

*Vulpes vulpes* Lineaus, 1758

Shaul Matsui (Hebrew)

Russeini, Abul hussein (Arabic), Biz Biz (colloquial Arabic)

**Description:** Head and body 520–660 mm.; ear 79–140 mm.; hind foot 114–155 mm.; tail 293–442 mm.

Although variable in color, the Common Red Fox may be separated into four geographical and ecological races in Israel, which probably intergrade. These are described separately below. It is distinguished by its fairly large size; pointed, blackish ears; and large, bushy tail, sometimes with a white tip. The upper parts, legs and tail range from pale to deep rufous or washed with black and either white, ashy or slaty throat and underparts. A black spot may be present mid-way between the eyes and the nose. The tail is usually more than half the length of the body. The cubs are gray-brown.

**Habits:** Mostly nocturnal and usually solitary, it lives in burrows in small family parties. Gestation is 52–53 days, and the litter may be up to seven cubs. It feeds mainly on small rodents, and sometimes fruit, carrion and grasshoppers.

**Habitat:** Rocky, forested mountains, coastal plains, stony desert wadis with some vegetation, not usually in sandy desert. It is sometimes found on the outskirts of towns and villages.

**Distribution:** Palearctic region. The Common Red Fox is ubiquitous in Israel. See subspecies for local distribution.

## Mountain Red Fox, Tawny Fox — Not illustrated

*Vulpes vulpes flavescens* Gray, 1843

**Description:** Head and body 465.2–572 mm.; ear 87–96.7 mm.; hind foot 120–134.7 mm.; tail 342–383 mm.

The long-haired Mountain Red Fox is usually a pale yellowish, stronger on the face and darker on the back. The sides of the forelegs and base of the tail are fulvous. Some individuals are more richly colored.

**Distribution:** Northern Iran, Kurdistan and Iraq. *V. v. flavescens* may be the subspecies found in the northern, more mountainous regions of Israel. Qumsiyeh (1996) claims *V. v. flavescens* is a synonym of *V. v. aegyptiacus*, but *V. v. aegyptiacus* is a junior synonym of *V. v. niloticus*.

## Palestine Red Fox — Plate 13

*Vulpes vulpes palaestina* Thomas, 1920

**Description:** Head and body 455–625 mm.; ear 83–105 mm.; hind foot 121–148 mm.; tail 305–412 mm.

The Palestine Red Fox is distinguished by its gray color, particularly along its sides, with a nearly complete suppression of rufous, except the face. The forelegs are grayish-rufous or fulvous. The underparts are whitish or black. The upper tail is buffy, washed with black.

**Distribution**: The subspecies *V. v. palaestina* is known from Lebanon and Israel, where it is common along the coastal plain and as far south as Beersheba.

## Egyptian Red Fox                     Plate 11

*Vulpes vulpes niloticus* Geoffroy, 1803

**Description**: Head and body 635 mm.

The Egyptian Red Fox is very slightly smaller than the Common Red Fox of Europe, paler with broader ears. It has variegated rufous or grayish-yellow upper parts, with a broad, rusty-red down along the middle of the back. Around the eyes and on the muzzle it is a rich rufous. The underparts vary from whitish to blackish. The limbs are rather rufous, with an irregular band down the front of the forelimb. The tail is rufous or rusty-yellow, sometimes washed with black. The black seems to increase with age.

**Distribution**: The race *V. v. niloticus* is known from Libya and Egypt. It may be the race that inhabits the mountains of the Negev and Sinai Deserts.

**Taxonomic Remarks**: The race *V. v. niloticus* is a senior synonym of *V. v. aegyptiacus* Sonnini, 1816.

## Arabian Red Fox                     Plate 13

*Vulpes vulpes arabica* Thomas, 1902

**Description**: Head and body 420–572 mm.; ear 73–109 mm.; hind foot 101–137.5 mm.; tail 258–383 mm.

The Arabian Red Fox is a small, pale subspecies with a suppression of black and relatively large ears. The upper parts are a pale rufous or buffy, variably speckled with white, gray and black. There is a tawny mid-dorsal stripe and grizzled gray on the posterior sides. The underparts vary from whitish to slaty-black or black. The limbs are pale buffy, fading to white distally on the paws. The tail is dull fulvous to buffy-white, the upper surface slightly stronger in color.

**Distribution**: Iraq, Saudi Arabia and Yemen. In Israel, the subspecies *V. v. arabica* is found in the southern half of the country, in the stony desert hills and wadis of the Negev Desert and the Arava Valley.

## Blanford's Fox                     Plate 13

*Vulpes cana* Blanford, 1877

Shual Tsukim (Hebrew)

Sha Rubah (Arabic)

**Description**: Head and body 406–438 mm.; ear 81–88 mm.; hind foot 92–101 mm.; tail 324–328 mm.

Blanford's Fox is distinctive for its fairly small size; very large, grayish-fawn colored ears; naked soles; delicate toes; and enormous black-tipped tail. The upper parts are gray, tinged with rufous on the crown, shoulders, along the back, tarsus and tail. A large black spot on each side of the narrow muzzle extends from the eye to the rhinarium. The underparts are white. Some specimens have a dark stifle.

**Habits:** It is strictly nocturnal. Extremely agile, it leaps from rock to rock with extraordinary speed and agility. The litter consists of 1–3 pups, born after a gestation period of 50–60 days. Its diet includes small mammals, birds, reptiles, arthropods and plants.

**Habitat:** Desert, on steep cliffs and mountain slopes strewn with boulders and little vegetation.

**Distribution:** Uzbek, southern Turkman, Russia, Afghanistan, Iran, northwestern Pakistan, Saudi Arabia, Oman, United Arab Emirates, Jordan, eastern Egypt (Sinai) and Israel. The author identified it as occurring in Saudi Arabia for the first time from a photograph taken by Mrs. Collenette on Jabal Shada (Gasperetti et al., 1985). In Israel, it was discovered by G. Ilani, where it is known from the western side of the Dead Sea (Ein Gedi), south to Eilat.

---

## Rueppel's Sand Fox                          Plate 11

*Vulpes rueppellii sabaea* Pocock, 1934
Shual HaNegev or Shual Holot (Hebrew)
Ta'alib, Husseine, Abul Hussein (Arabic),
Biz Biz (colloquial Arabic)

**Description:** Head and body 342–450 mm.; ear 88–110 mm.; hind foot 90–115 mm.; tail 260–355 mm.

In Israel, Rueppel's Sand Fox is distinguished by its relatively small size; pallid color; very large yellowish-fawn ears, without black; and white-tipped tail. The upper parts and tail are pale sandy-buff; very little buffy-orange

around the eyes and shoulders; a light, orange-rufous mid-dorsal stripe; and pale, buffy-yellow flanks, with little gray. The blackish patch on the sides of the muzzle before the eye is weak or lacking. There is a blackish glandular spot on the upper surface of the tail. The underparts and feet are whitish. The soles of the feet are densely furred. *Vulpes rueppellii sabaea* is distinguished from the African subspecies *V. r. reuppellii* Schinz, 1825 by its paler color. A specimen from the southern Arava Valley (Ein Yahav) in Israel was extremely pallid, lacking any buffy-orange on the face, or black on the muzzle. It clearly resembles the Arabian subspecies *V. r. sabaea*. The nominate subspecies is more richly colored, grayer with more buffy-orange around the eyes and forehead. The newborn pup is grayish-brown.

**Habits:** It is mostly nocturnal and feeds on small wildlife, i.e. jerboas, lizards, centipedes and probably some plants. It also scavenges at garbage dumps. Gestation is 56 days, and the litter 1–4.

**Habitat:** Deserts and steppes, in low stony hills and wadis, in deep burrows or caves.

**Distribution:** North Africa, from Algeria, Libya and Egypt, south to Sudan, Somaliland and Asben, Iran and Afghanistan. The subspecies *V. r. sabaea* is known from Iraq, Saudi Arabia and Yemen. In Israel, it inhabits the western side of the Dead Sea, and Arava Valley. It may intergrade with the African subspecies, *V. r. rueppellii*, in the Negev and Sinai Deserts where intermediate forms occur.

# BEARS: URSIDAE

Bears are among the largest living creatures. They are characterized by their burly proportions, with thick legs; powerful forelimbs; a long muzzle; small, rounded ears; and a short tail. They have five toes on both front and hind feet, with large nonretractile claws. Mainly terrestrial, they walk with a slow and measured gait on the soles of their feet. Their fur is usually long, coarse and shaggy, distinguished by the dominance of black, although there are several brown and one white species. Widespread in distribution, they are found mostly in the northern hemisphere. One species inhabits South America, and another reaches North Africa. A single species, now extirpated, formerly inhabited the northern part of Israel.

**Syrian Brown Bear**          **Plate 14**

*Ursus arctos syriacus* Hemprich and Ehrenberg, 1828

Dov Suri (Hebrew)

Dub (Arabic)

**Description**: Head and body 1,535 mm.; tail 35 mm.

The Syrian Brown Bear is a southern subspecies of the Eurasian Brown Bear. It is characterized by its relatively small size; pale straw color, sometimes fulvous or nearly brown; and long white claws.

**Habits**: It is solitary and unsociable. There is delayed implatation from 4.5 to 7 months. Effective gestation is only about 60 days, but with delayed implantation it is 210–250 days. The litter is 1–5, often twins, born during winter "hibernation", in a natural shelter or a den dug in a hillside. The mother cares for the young for a year or more. Although omnivorous, including carrion, it usually feeds on plant matter.

**Habitat**: Mountain steppes, arid, wooded hills, often at high altitudes.

**Distribution**: The species ranges widely across the northern parts of the New and Old Worlds. The subspecies *U. a. syriacus* is known from Asia Minor, Caucasus, Iran, Iraq, Syria and Lebanon. In Israel, it formerly occurred in Galilee and the Judean hills during biblical times, but was extirpated about a hundred years ago. The last wild Syrian Bear was killed near Majdal Shams on Mount Hermon. Today, it exists in Israel only in zoos.

# WEASELS, POLECATS, MARTENS, BADGERS AND OTTERS: MUSTELIDAE

## WEASELS, POLECATS AND MARTENS: MUSTELINAE

Martens and Polecats are moderately small carnivores, distinctive for their long, slender bodies and small, sharp, "pushed-in" faces, with short, rounded ears; short legs; and relatively bushy tails. They vary greatly in color, some being uniformly brown or white, whereas others conspicuously marked in bold patterns. They have five toes on each foot and short claws that may be partly withdrawn. The bottom of their feet are furred, and they walk on their toes. Most members of the family emit a noxious-smelling substance from anal scent glands. Almost exclusively cosmopolitan, they occur in virtually every type of terrestrial habitat, from the Arctic tundra to tropical rain forests. In Israel, they are represented by five or six species of mustelids.

### Common, Least or Snow Weasel     Plate 15

*Mustela nivalis* Linnaeus, 1766

Hamoos HaShlagim or Hamoos Gamadi (Hebrew)

**Description**: Head and body 160–290 mm.; hind foot 20.5–30.5 mm.; tail 40–70 mm.

The Snow Weasel is the smallest carnivore in the region. It is distinguished by its slender body; long neck; low, rounded ears; short limbs; and tail which is less than a quarter of the length of the head and body. In the summer, the upper parts are a uniform brown, and the underparts are white, sharply demarcated along the flanks. The dorsal surface of the forefeet is white. The tail is brown, becoming darker towards the tip. The winter coat is presumably all white, as in the colder parts of its range.

**Habits**: It is active day and night. It inhabits holes, often the burrows of rodents and hollow trees, among boulders and rock crevices. It feeds on small rodents, birds, lizards, amphibia and occasionally larger animals. Gestation is 34–37 days, and it usually produces 3–9 young.

**Habitat**: Mountains, as high as the subalpine zone.

**Distribution**: Widespread in Europe eastwards through Russia, Asia Minor, Iran, northern Arabia, Afghanistan, Mongolia, Korea, China, Japan and North Africa, Egypt, Morocco, Algeria and North America. The subspecies found in Lebanon is smaller than the Egyptian subspecies *M. n. subpalmata* and may belong with the Mediterranean subspecies, *M. n. boccamela*. Although reported from the vicinity

79

of Mount Tabor, it is rare and little known in the region and its occurrence in Israel needs confirmation.

## Beech Marten, Stone Marten          Plate 17

*Martes foina syriaca* Nehring, 1902
Dalak (Hebrew)
Sinsar (Arabic)

**Description**: Head and body 407–446 mm.; ear 36.9–41 mm.; hind foot 75.2–85 mm.; tail 186–240 mm.

The Beech Marten is of medium size with a slender body; short, rounded ears; short legs; and a long, bushy tail. It is distinguished by its variable light or dark brown head and body, sharply contrasting creamy-white throat and breast, interrupted by some dark spots, dark brown limbs and tail.

**Habits**: It is both diurnal and nocturnal. A swift and agile climber, it is usually arboreal, although sometimes it is also terrestrial. Delayed implantation is 230–275 days. Effective gestation is 30 days, and the litter is usually 3–4. Omnivorous, it preys on small wildlife and eats fruit.

It's voice is a metallic squeaking, a repeated "Chuck, chuck", growling or squealing.

**Habitat**: Wooded areas, particularly evergreens, as well as barren, stony hills and mountains.

**Distribution**: Widespread across Europe, Asia Minor and Asia. The subspecies *M. f. syriaca*

occurs in Iraq, Syria, Lebanon and Jordan. In Israel, it was formerly common in the Judean hills and was extirpated on Mount Carmel. It has recently appeared at Ramat Shaul and Kiryat Shprinzak. It is now rare in the Galilee and the Golan Heights, but has increased in the Hula valley near Kibbutz Dan.

## Syrian Marbled Polecat          Plate 16, Figure 7

*Vormela peregusna syriaca* Gulddenstaedt, 1770
Samoor (Hebrew)
Abulfiss (Arabic)

**Description**: Head and body 275–299 mm.; ear 16.7–27 mm.; hind foot 32.8–39.3 mm.; tail 175–178 mm.

The Marbled Polecat is the most colorful mammal in Israel and the only member of its genus. It is distinguished by its fairly small size; slender body; and short legs. This race is characterized by its brighter coloration, a strong pattern of light yellow-orange upper parts with brown spots; white forehead and cheeks; a broad, blackish facial mask; and a blackish band across the ears and crown. The occipital region is variable, white with spots or streaks. The underparts are blackish-brown, somewhat rufous above the foreleg and blackish limbs. The long, bushy tail is a mixture of brown, yellow and black.

**Habits**: Usually nocturnal and solitary, it excavates burrows, as well as using those of other animals. It is mainly terrestrial but can also climb well. Gestation is 23–45 days, but with delayed implantation from 6–9 months for a total of 8–11 months. The litter size is 1–8. It preys on rodents, birds, reptiles and sometimes poultry. When threatened, it curves its tail over its back as a warning.

**Habitat**: Open areas on steppes, hills and plains.

**Distribution**: Ranges from southeastern Europe and southwestern Asia, Russia into Mongolia. The subspecies *V. p. syriaca* is found from Syria to western and northern Iraq. In Israel, it is fairly common in the northern half of the country up to the edge of the desert.

## BADGERS AND OTTERS: MELINAE

Badgers and otters are medium-sized, short-legged carnivores, differing greatly in proportions and color. The Badger is stocky with a short tail, whereas the Otter is slender and has a long tail. The Badger is strictly terrestrial, whereas the Otter is mainly aquatic. Badgers and otters are found in both the New and Old Worlds. The Badger is confined to the northern hemisphere, the Ratel in Africa and Asia, while the Otter is found throughout North America, Europe, Asia Minor, Asia, Arabia and Africa. In Israel, they are each represented by a single species.

### Persian Badger, Old World Badger, Eurasian Badger

Plate 18

*Meles meles canescens* Blanford, 1875
Gireet Metsuya (Hebrew)
Ghreir or Anak U'lord (Arabic)

**Description**: Head and body 600–650 mm.; ear 30–55 mm.; hind foot 95–150 mm.; tail 100–155 mm.

The Persian Badger is fairly large, stocky, with short, rounded ears, legs and tail and has a distinctive facial pattern. The head is white, with a broad black stripe on each side of the face, beginning between the nose and eye and extending backwards to include the ears. The upper parts and tail are a grizzled gray, more or less tinged with rufous, with blackish underparts and legs. The front feet have long claws.

**Habits**: Mostly nocturnal and fossorial, it burrows in large mounds of earth with several entrances, called a set. Delayed implantation of 3–10 months includes 49 days gestation. There are usually 2–7 young. It is omnivorous, feeding on both plants, roots, small animals, insects and carrion. In the north, it hibernates, but in Israel it remains active during the winter.

**Habitat**: Woods, open lands and vineyards.

**Distribution**: The only species of its genus, it is widespread throughout Europe and Asia, Tibet, northern Burma and southern China. The race *M. m. canescens* occurs in Iran, Syria and Lebanon. In Israel, it is uncommon but has been recorded in Upper Galilee, Jezreel valley, upper Jordan Valley and the coastal plain.

## Honey Badger or Ratel          Plate 18

*Mellivora capensis wilsoni* Cheesman, 1920
Gireet Ha'dvash (Hebrew)
Abu Djaga, Abu Keem, Rorejri, Ghreir, Dhirban or Dhrambul (Arabic)

**Description**: Head and body 505–737 mm.; ear 27–35.4 mm.; hind foot 78–108 mm.; tail 147–246 mm.

The Honey Badger is of medium size and characterized by its stout build; short, rounded ears; short legs; short, bushy tail; and large claws on the forefeet. Its coloration is contrary to most mammals; it has light gray upper parts which become yellowish-white on the crown, while the face, underparts, limbs and tail are dark brown or black.

**Habits**: It is generally nocturnal, but may come out during the day. Mostly terrestrial, it occasionally climbs a low-slung tree. It feeds mostly on insects, small mammals, lizards and snakes, including venomous ones, such as cobras. It also eats carrion, plants and is fond of honey. In Africa, it has a remarkable symbiotic relationship with the Honey Guide, a bird that, upon finding a bee's nest, leads the Honey Badger to it. Impervious to the bee's stings, the Honey Badger tears open the nest for the honey, and the Honey Guide is rewarded with honeycombs. The Honey Badger is courageous and will attack animals much larger than itself in defense. After a gestation period of six months, usually two young are born.

**Habitat**: Variable, from tropical to arid regions. It inhabits hollow trees, burrows, or rocky ravines.

**Distribution**: Most of Africa, Arabia to Russian Turkestan, east to Nepal and India. The subspecies *M. c. wilsoni* is known from Iran, Iraq, Syria and southern Arabia. In Israel, it is rare but widespread from Upper Galilee (Umm Falik) to the Judean hills and the Negev Desert (Ein Hussub). It has also been recorded from Gaza.

## Common Otter, River Otter          Plate 15

*Lutra lutra seistanica* Birula, 1912
Lutra (Hebrew)
Kelb el ma (Arabic)

**Description**: Head and body 590–960 mm.; ear 19.3–30 mm.; hind foot 120–145 mm.; tail 360–550 mm.

The Common Otter of the subfamily Lutrinae, is distinguished by its flattened head, broad muzzle; small ears; elongated body; short legs, with webbed toes; and long, thick, tapering tail. The Middle Eastern race is paler than the otter of Eurasia. The short, soft, glossy fur is a uniform light grayish-brown above, although it is variable. The underparts are grayish-white.

**Habits**: Amphibious, it is seldom found far from water. The small ears and nostrils can be closed when submerged. Active day and night, it is an excellent swimmer and diver. It is usually solitary, but is sometimes found in pairs. It is among the most playful of mammals. Its den is a burrow in the side of a bank. After a gestation period of about 60 days, a litter of 1–5 is born. They remain together for about eight months. It feeds on fish, crayfish, frogs, turtles, small mammals, and birds are also taken.

**Habitat**: Lakes, ponds and perennial rivers.

**Distribution**: Widespread across Europe and Asia, from England to Japan, Asia Minor, Arabia and North Africa. In Israel, the subspecies *L. l. seistanica* is widespread, though uncommon, in the northern half of the country, from the Huleh Valley to the mouth of the Jordan at the Dead Sea and the coastal plain.

# MONGOOSES, CIVETS AND GENETS: VIVERRIDAE

## MONGOOSES: HERPESTINAE

The mongoose is a small- to medium-sized carnivore, with an elongated body and long tapering tail. The muzzle is small and pointed, and the ears are short and rounded. The pupil of the eye is horizontal. The legs are short, with five toes on each foot, and a very small inner toe. The claws cannot be withdrawn. It is generally uniformly colored, although the individual hairs are speckled alternately with dark and light rings, which give the coarse fur a grizzled appearance. It is mainly terrestrial and shelters in holes. It is one of the few species of mammals that are active during the day and early evening. Widespread throughout Africa and Asia, they are represented in Israel by a single species.

The Palestine Genet, *Genneta genetta terraesanctae* Neuman, 1902, was recorded from Mount Carmel by Tristram (1866). Its actual provenance is not certain: whether it was collected on Mount Carmel or obtained from some other source. Unspecified additional specimens were reported from Sejera and Wadi Fauar near the Dead Sea by Aharoni (1912). The existence of these specimens is unknown. Geographically, the nearest the genet is known to Israel is in Upper Egypt and Yemen, yet the subspecies *G. genetta terraesanctae* is closer morphologically to the genet of western Europe (Harrison, 1968). The presence of the genet in Israel is, therefore, in doubt.

## Egyptian Mongoose, Ichneumon Plate 15

*Herpestes ichneumon ichneumon* Linnaeaus, 1758

Nemeah (Hebrew)

Zerdi or Abul Irris (Arabic)

**Description**: Head and body 510 mm.; ear 22–25 mm.; hind foot 80.5–90.5 mm.; tail 330–450 mm.

The Egyptian Mongoose is of medium size (large for a mongoose), with an elongated, low-slung body; short, broad ears; relatively short legs; and a long, tapering tail. It is distinctive for its uniformly-grizzled, brownish-gray body and tail, speckled with black and white. The head and limbs are darker, and the tail terminates in a long, black, brush-like tassel.

**Habits**: Mostly nocturnal, but also diurnal, it appears singly, in pairs, or in a family following each other in a line. It is completely terrestrial, inhabiting burrows or crevices under rocks. A litter of 2–4 is born in a burrow or in a hollow tree after a gestation of about 60 days. The young remain with the mother after the birth of a subsequent litter. Its diet consists of small wildlife, domestic animals, cobras and other venomous snakes, as it is resistant to their poison. An inverse relation has been found between mongoose density and the incidence of Palestine Vipers in Israel.

**Habitat**: Open lands among shrubs and rocks, near water, along hedges, orange groves, around cultivated lands, fish ponds, agricul-tural settlements and on the outskirts of villages.

**Distribution**: Southern Spain, North, East and Southwest Africa, Asia Minor and Turkey. In Israel, the subspecies *H. i. ichneumon* is common in the northern half of the country, in the Huleh Valley, along the coastal plain, with several isolated populations near the Dead Sea and in the Arava Valley.

# HYAENAS: HYAENIDAE

Hyaenas are fairly large, dog-like animals, but they have longer forelegs than hind legs which gives them a distinctly front-heavy, ungainly appearance. They have short, broad heads with blunt muzzles and large, broad, pointed ears. Their powerful jaws and dentition are used for crushing bones. Their feet have only four toes each and are furnished with short, blunt, nonretractile claws. Their fur is coarse and shaggy, with an erectile dorsal crest, and more or less marked with either stripes or spots. The family is represented by only one genus. Confined to the warmer parts of the Old World, they have been extripated from Europe to the Bay of Bengal and China. In Israel, only one of two native species survives.

## Syrian Striped Hyaena    Plate 19

*Hyaena hyaena syriaca* Matschie, 1900

Tsavo'a Mefuspas (Hebrew)

Dhaba or Dab (Arabic)

**Description:** Head and body 975–1240 mm.; ear 150 mm.; hind foot 220 mm.; tail 295 mm.

The Syrian Striped Hyaena is characterized by its large size and ungainly proportions. The head is massive; the ears are long, broad and pointed; and the back is sloping. The tail is long and bushy. In the winter, the coat is long, and a shaggy, erectile mane partly obscures a blackish dorsal stripe. The general color is a pale grayish-white, with some buffy tint on the crown and tail. The face, muzzle and throat are dark. A black bar extends up the cheek towards the ear; another smaller one lies between it and the eye; and several black spots above it are more or less distinct. The flanks and shoulders are vertically striped with black, and the legs with transverse black stripes. The underparts are a dirty white, except for the throat, which is blackish.

**Habits:** Nocturnal, it seldom appears during the day. It is solitary, and rarely more than two are found together. It is shy and retiring, not aggressive, and will flee rather than fight. Although it can run quickly, its pace seems slow and ungainly, owning to the large front legs, and it runs with a lumbering gait. It is usually silent but sometimes emits a sort of laughing cry. After a gestation period of about 90 days, 2–6 young are born and raised in a hole in the ground or in a cave. It is primarily a scavenger, but will take small animals. It sometimes forages at night around human habitation for edible refuse. It carries food back to its lair in caves or earth dens which are littered with bones.

**Habitat:** Open rocky hills, arid savannah, often not far from cultivated land.

**Distribution:** North and East Africa, Egypt and

Sinai, through Asia Minor, southern Russia, Iran, Arabia, Lebanon, Syria, Jordan, Iraq to Nepal and India. The subspecies *H. h. syriaca* is known from Syria, Lebanon and Jordan. In Israel, it has disappeared from the coastal plain and is becoming rare in the Huleh Valley, Upper Galilee, Mount Carmel and the Judean hills, south to the Negev Desert and the Arava Valley.

## Arabian Striped Hyaena                    Plate 19

*Hyaena hyaena sultana* Pocock, 1934
Tsavo'a Aravi (Hebrew)
Dhaba or Dab (Arabic)

**Description**: Measurements are not available. It averages smaller than the Syrian subspecies.

The Arabian Striped Hyaena is a desert subspecies, characterized by its smaller size on average and richer coloring. It is described as gray to whitish-gray, buffy or yellowish-tawny on the flanks, thighs and shoulders; darker and browner on the foreshoulder and neck up against the mane. The head is decidedly brown, buffy-yellow below the eyes, with a dusky-gray muzzle. The throat is blackish, with a black stripe extending up the side of the jaw towards the tall, narrow ears. Other facial spots are reduced; the stripes on the forelegs are broken into spots; and the tail mostly whitish. The young are more richly colored.

**Habits**: Nocturnal and solitary, it lives in caves or earth dens. Omnivorous, it is primarily a scavenger, but also catches live food. Sometimes, it enters human settlements. Gestation is 90 days and the litter 1–4.

**Habitat**: Desert fringes and garbage dumps.

**Distribution**: North and East Africa, through Asia Minor, southern Russia, Iran, Arabia, Lebanon, Syria, Jordan, Iraq to Nepal and India. The subspecies *H. h. sultana* is known from southern Arabia. In Israel, it occurs near the southern end of the Dead Sea (Neot Hakikar). A specimen in the collection of the Hebrew University of Jerusalem constitutes the first geographical record for Israel. It may be that the Arabian race intergrades with the Syrian subspecies in the northern part of its range.

# CATS: FELIDAE

Cats are small to large carnivores which, except for the cheetah, are quite uniform and resemble the domestic cat in proportions. They are easily distinguished from the dog family by their short, blunt muzzles; loose fur about their stomachs; and long, cylindrical tails. They vary in size more than any other carnivore. The pupils of their eyes contract vertically, except for the caracal. They walk upon their toes, which number five on each forefoot and four on each hind foot. The forelimbs are strongly built with sharp, retractile claws. Among the carnivores, they are the most proficient killers, taking prey as large or larger than themselves. They are generally marked, more or less, with spots and stripes, which help conceal them, although some are more uniformly colored. Cats are cosmopolitan, and were once represented in Israel by six species, two of which were extirpated and one which is in danger of extirpation.

## Palestine Wild Cat, Bush Cat Plate 20, Fig. 15

*Felis silvestris tristrami* Pocock, 1955

Hatoul Bar (Hebrew)

Kot el Khla (Arabic)

**Description**: Head and body 406–498 mm.; ear 55–58 mm.; hind foot 115–131 mm.; tail 282–390 mm.

The Palestine Wild Cat is a local subspecies of the wildcat of Eurasia and Africa, and the ancestor of the domestic cat. It is rather small and distinguished by its generally light grayish-buffy or yellowish-gray upper parts; with indistinct dark vertical stripes or spots; two distinct facial stripes, extending backwards from the eye and cheek; and paler underparts. The ears are small and rufous on their backs, with a faint black tip. The upper limbs are marked with broad, dark bands. The inner side of the upper forelegs have a distinct black band, and the soles of the feet are black. The tail is relatively long, concolorous with the back, but with several blackish rings and a black tip. It may be confused with a juvenile Jungle Cat.

**Habits**: Usually nocturnal, it is also often active during the day. Otherwise, its habits are similar to those of the domestic cat, although it has a wilder disposition. It sometimes interbreeds with domestic cats and is difficult to distinguish. It shelters in hollow logs, crevices, abandoned burrows and vegetation. Gestation is 63–66 days, and the litter consists of 1–6 kittens. Its diet consists of small animals.

**Habitat**: Steppes, hills, valleys and the coastal plain, in open rocky areas as well as forests and sand dunes.

**Distribution**: The species is widespread in Europe, Asia, Arabia and Africa. The subspecies *Felis silvestris tristrami* is found in

Lebanon, Syria and Jordan. In Israel, it is fairly common throughout most of the country.

## Desert Wild Cat                                Plate 20

*Felis silvestris iraki* Cheesman, 1921
Hatoul Bar (Hebrew)
Kot el Khla (Arabic)

**Description:** About the same size as the Palestine Wild Cat.

The Desert Wild Cat is a little-known desert subspecies of the wild cat of Eurasia and Africa. It is distinctive for its pale, sandy-gray upper parts; indistinct spinal band; and all the markings on the flanks and legs are smaller and weaker. The facial stripes and backs of the ears are pale rufous. There is an incomplete darker band on the upper, inner side of the foreleg. The underparts and feet are whitish. The soles of the feet are black and somewhat hairy. The blackish rings on the tail are incomplete, and the tip is black.

**Habits:** Unknown. Probably similar to the Palestine Bush Cat.

**Habitat:** Desert. The pale color and slightly hairy soles suggest it may be adapted to a more sandy desert than *F. c. tristrami*.

**Distribution:** Widespread in Europe, Asia and Africa. The race *F. c. iraki* was described from Kuwait and northeast Arabia. In Israel, a specimen fitting the description of this race, which had been killed by a car, was found by the author on the western side of the Dead Sea between Ein Zohar and Ein Boqek.

## Sand Cat                                         Plate 21

*Felis margarita* Loche, 1858
Hatoul Holot (Hebrew)
Kot (Arabic)

**Description:** Head and body 390–550 mm.; ear 57–68 mm.; hind foot 110 mm.; tail 230–340 mm.

The Sand Cat is distinctive for its small size; pale color; and broad face with large, outstanding ears, which make the head appear flat. The upper parts are uniformly pale sandy-buff, darker on the back, with inconspicuous markings. It has a whitish face, with a rufous streak extending back from the eye, and a black spot on the backs of the ears. Several dark brownish stripes cross the upper parts of the legs. The soles of the feet are covered with long, grayish hairs. The tail has a few black rings and a black tip. The juvenile has stronger markings than the adult.

**Habits:** It is nocturnal and inhabits sand dunes and sometimes rocky areas, where it lives in shallow burrows under scrub. Gestation is 63 days, and the litter 2–5. Its diet contains small animals, and it can exist on a minimum of water.

**Habitat:** Sandy desert, also rocky terrain.

**Distribution:** North Africa, Egypt (Sinai), Russian Turkestan and Arabia. In Israel, it is confined to the Arava Valley (Hatseva).

**Taxonomic Remarks:** The Sand Cat in Israel has been described as the subspecies *Felis margarita harrisoni* Grubb and Groves, 1976, based on four specimens. Subspecies of Sand Cats have been based on very few specimens and their variation is very uncertain (Harrison, 1968).

## Palestine Jungle Cat, Swamp Cat   Plate 22

*Felix chaus furax* de Winton, 1898
Hatoul Bitsot (Hebrew)
Bizoon, Bizoon el Berr, Kot-wah Shee, Kot Buri, Asad, Sabua or Lebuah (Arabic)

**Description**: Head and body 487–777 mm.; ear 71–73.6 mm.; hind foot 147–168 mm.; tail 247–272 mm.

A slightly larger subspecies of the Jungle Cat of North Africa and southwest Asia, it is distinguished by its tufted ears; long legs; relatively short tail; and unmarked, grizzled, yellowish-gray or gray-brown upper parts. It is darker on the back and tinged with rufous. Its face is lighter with a brown lacrymal stripe, and the cheeks may be faintly banded or pale. The ears are pale rufous, with a large, black tip and apical tufts. The long legs may be transversed by dusky bars. The inner side of the upper foreleg has two broad, blackish transverse bars. The underparts are whitish, tinged with yellow or rufous. The soles of the feet are dark brown. The tail is grayer than the back and has two black rings and a black tip. The juvenile resembles a wild cat, but can be distinguished by its juvenile proportions, ear tufts and relatively shorter tail.

**Habits**: It is nocturnal. Its den is made of dry vegetation or burrows of other mammals.

Gestation is 66 days, and the litter is usually 2–3. It preys on small wildlife and sometimes poultry. Often seen in pairs, it may be monogamous.

**Habitat**: Marshes, papyrus swamps, reed beds, fields of tall grass, such as corn or sugar cane.

**Distribution**: Asia, from the Caucasus and Turkestan to India and the Indochinese pensinsula, and Egypt. The subspecies *F. c. furax* is known from Iraq and Jordan. In Israel, it is found in the Huleh and Jordan Valleys, Galilee, the coastal plain, reaching just north of Beersheba, Jericho and the southern end of the Dead Sea.

## Arabian Caracal,   Plate 23, Cover
## Desert Lynx

*Lynx caracal schmitzi* Matschie, 1912
Hatoul Midbar or Karakal (Hebrew)
Itfah, Niss or Anag el Ard (Arabic)

**Description**: Head and body 394–630 mm.; ear 63.5–74 mm.; hind foot 101.5–157 mm.; tail 152–227 mm.

The Arabian Caracal is a fairly large subspecies of the Afro-Asian Caracal. It is robust, with a flat head and long legs, the hind legs longer than the front ones. It is distinguished by its uniformly tawny-rufous upper parts, with a vinaceous tinge; the black backs of the ears which are surmounted by long, black tufts; and its relatively short tail. The underparts are whitish with some obscure spots, more distinct in the juvenile. There is a rare gray phase. The young are brighter in color.

**Habits:** Mostly nocturnal, it retreats to boulders during the day. Gestation is on average 78 days. A litter of 1–5 is hidden in crevices or hollow trees. It feeds on small mammals, birds and lizards and sometimes small livestock and poultry. It is particularly good at catching birds while they are in flight.

**Habitat:** Hilly country with boulders, steppes and desert.

**Distribution:** Northern Africa, Arabia, the Near East and India. The subspecies *L. c. schmitzi* is known from Iraq, Syria, Lebanon, Jordan, Saudi Arabia, Kuwait and Oman. In Israel, it has been found in the Golan Heights, Upper Galilee, the Jordan Valley, Mount Carmel, near the Dead Sea, in the Negev Desert and Arava Valley, south to Eilat.

**Taxonomic Remarks:** The caracal was originally placed in the genus *Felis* L, 1758. The genus *Lynx* L, 1758, was described at the same time. It was later put in the subgenus *Caracal* Gray, 1843. It is readily distinguishable from the genus *Felis*, and essentially the same as the *Lynx*, in its skull, dentition and external habitus (Harrison, 1968). Its pupil does not contract to a circle in dim light but rather to a vertical slit as in other smaller cats (MacDonald, 1984). There does not seem to be sufficient reason to recognize *Caracal* as a valid subgenus.

## Persian Lion, Asiatic Lion — Plate 26

*Panthera leo persicus* Meyer, 1862
Ari, Aryeh, Levi, Schachal (Hebrew)
Saba or Seba, Asced (Arabic)

**Description:** The largest recorded measurement of an Asiatic Lion is 3,230 mm.

The Asiatic subspecies of the lion is distinguished from the African Lion in that it is smaller and has a heavier coat of hair, but a less-prominent mane. The belly fringes and elbow tufts are more pronounced and may be black. The tail has a longer tassel. The female does not have a mane. The young are partly spotted.

**Habits:** Diurnal and nocturnal, it is sociable and lives in family groups (prides), seldom alone. It has a wide home range. The gestation period is 105 days, and there are 2–6 cubs per litter. It is one of the most powerful carnivores and feeds on certain preferred ungulates, but may take smaller prey, cattle and even man. Its roar is well known.

**Habitat:** Savannahs, forests and thickets near a river.

**Distribution:** Formerly Greece, Asia Minor, Iran, northern Arabia east to India, where a few hundred survive. The subspecies *P. l. persicus* was found in Iran, Iraq and Syria. In Israel, it was known in biblical times. The last one is said to have been hunted near Meggido in the thirteenth century. It survived in Mesopotamia until the nineteenth century. The lion has not been reported from Iran since 1942. However, it is possible that it still exists there.

## Syrian Leopard                 Plate 24

*Panthera pardus tulliana* Valenciennes, 1856

Namer Suri (Hebrew)

Nimir, Faha (Arabic)

**Description**: Head and body approximately 1,310–1,540 mm; tail 750–1,090 mm.

The Syrian Leopard is a northern subspecies of the Afro-Asian Leopard. It is distinguished from the Arabian subspecies by its larger size and longer coat, particularly in the winter. It also has darker upper parts; a darker mid-dorsal region, with large rosettes that are thin-rimmed and more widely spaced; and whitish belly. The tail is long and spotted, and appears to be bushier.

**Habits**: Nocturnal and crepuscular, it is elusive and solitary except during mating season. It hides in thick bushes, rocks or caves. It climbs trees where it often carries its prey. After a gestation period of about 91 days, 1–3 cubs are born, but usually do not all survive. Its diet consists of wild animals and sometimes domestic animals.

**Habitat**: Rocky hills, steppes and mountains.

**Distribution**: Widespread from South Africa to Iran, Arabia and Asia, as far east as Japan. The subspecies *P. p. tulliana* is known from Turkey, Syria and Lebanon. In Israel, it has been recorded from Upper Galilee, formerly Mount Carmel, and the Judean hills (near Jerusalem).

## Arabian Leopard, Sinai Leopard   Plate 25

*Panthera pardus nimr* Hemprich and Ehrenberg, 1833

Namer (Hebrew)

Nimir, Faha (Arabic)

**Description**: Head and body 912–1,140 mm.; ear 46–76 mm.; hind foot 210–250 mm.; tail 730–880 mm.

The Arabian Leopard is a small desert subspecies of the widespread Afro-Asian Leopard. It is distinguished by its relatively large size; shorter pelage; pale buff or yellowish-tawny upper parts, suffused with buffy-brown, including the underparts and legs, with very little contrast between the flanks and belly; and is spotted with large rosettes of brown, their centers slightly darkened. The back of the ears are black with a white median spot. The tail is long and spotted. Females are similar to the males but usually smaller. The spots on cubs are rather indistinct.

**Habits**: Similar to the Syrian Leopard. It preys on birds and mammals, especially the hyrax, ibex and porcupines, and may take domestic animals. A male's territory is very large and could include the territories of 2–4 females.

**Habitat**: Rocky desert hills and mountains, not far from water.

**Distribution**: Widespread from South Africa to Iran, Arabia, and Asia as far as Japan. The Arabian subspecies *P. p. nimr* occurs in Israel along the western side of the Dead Sea (Ein Gedi), the Judean Desert and Negev Deserts, south to Eilat and Sinai. It is rare and on the verge of extirpation.

## Asiatic Cheetah Plate 26

*Acionyx jubatus venaticus* Griffith, 1821
Bardelas Asiati (Hebrew)
Fahed, Fah'ad (Arabic)

**Description**: Head and body 1,400–1,500 mm.; tail 600–750 mm.

The measurements of the subspecies *Acinonyx jubatus venaticus* are unknown except for a skin which measures: head and body 1,270 mm.; tail 685 mm.

The Asiatic Cheetah is large and slender, with a small head, short ears and long, thin legs and tail. Its general color is tawny-buff, paler on the flanks and merging into whitish underparts. A distinct black stripe runs from the eye to the mouth. The coat is completely covered with small, evenly-scattered, solid, black spots, which are less distinct on the underparts. There is a slight wiry mane on the neck and shoulders. The claws are semi-retractile. The long tail is spotted, with bands near the white tip. Cubs are smoky-gray at first, without markings.

**Habits**: It is solitary or found in numbers up to six. It is active in the mornings and evenings, sometimes on moonlit nights. It is the fastest cursorial animal in the world and can sprint up to 96 km. per hour. The gestation period is about 95 days, and the litter 2–4 cubs. Many do not survive. Unlike other large cats, it purrs like a house cat, only louder. It preys on smaller and young mammals, and birds by running them down.

**Habitat**: Open plains, semi-deserts and steppes.

**Distribution**: Africa, Asia Minor and southwest Asia. The Asiatic Cheetah has not been seen in India since 1948 and is apparently extirpated except from Iran and possibly Afghanistan. By 1884, it was scarce in Palestine, though more common east of the Jordan River. By 1930, it was rare, but still common in the southern steppe. In Israel, the last one was seen in the Negev Desert (near Yotvata) in 1959.

# UNGULATES OR HOOFED MAMMALS: UNGULATA

Ungulates are mostly large-hoofed mammals and include the only mammals with horns or antlers. They are generally divided according to whether they have an odd or an even number of toes. They are almost completely herbivorous.

# ODD-TOED UNGULATES: PERISSODACTYLA

Odd-toed ungulates are medium to large herbivorous mammals, characterized by the fact that with rare exception, they have one or three digits on their toes. In Israel, they are represented by the hyrax.

The Syrian Onager or Wild Ass, *Equus hemionus hemippus*, the smallest wild horse inhabited the Syrian desert where it became extinct in the 1920s. Although it is not native to Israel in historic times, a hybrid population between the larger Persian Onager, *E. h. onager*, and the Asiatic Onager, *E. h. kulan*, has been introduced into the central Negev Desert.

Fig. 8. Sinai Rock Hyrax

# HYRAXES: HYRACOIDEA

## HYRAXES: PROCAVIA

Hyraxes are the smallest hoofed mammals. The hyrax superficially resembles a rodent, but is actually descended from the same ancestors as the elephant, its nearest living relative. The hyrax is characterized by its stocky appearance; short, rounded ears; and no visible tail. It has four toes on each front foot and only three toes on each back foot. The innermost hind toe bears a curved nail. On the soles of the plantigrade feet are specialized, elastic pads that function as suction cups for traction. The hyrax has a wide geographical and altitudinal distribution. It is found in Africa and Arabia. A single species inhabits Israel.

### Syrian Rock Hyrax                    Plate 27

*Procavia capensis syriacus* Schreber, 1784
Shafan Suri (Hebrew)
Tubsun, Webr or Webeh (Arabic)

**Description**: Head and body 450–574 mm.; ear 31–40 mm.; hind foot 77–81 mm.

The Syrian Rock Hyrax is of medium size and variable in color, from light to dark brown, with a light brown dorsal spot. It is distinguished by its stocky shape, with small rounded ears; short legs; and no tail.

**Habits**: Active during the day, it can be seen sitting, sprawled out on a rock or nimbly jumping from rock to rock. It sometimes climbs a tree like its cousin, the Tree Hyrax. The hyrax is usually difficult to approach as a sentry is posted to warn the band of danger by a peculiar, shrill cry. The hyrax apparently has a low metabolic rate and poor ability to regulate its body temperature. This is maintained by basking, huddling together, and being inactive for relatively short periods. Polygamous, they live in cohesive and stable family groups of one male and 3–7 related females. The gestation is 210–240 days with up to four young. The Syrian Rock Hyrax is preyed on by leopards, jackals and eagles, especially the Black Eagle, which used to live in Israel. It feeds on a variety of plants, but mainly grass, and is able to eat the poisonous oleander with impunity. The Syrian Rock Hyrax was mistakenly considered a ruminant on account of the way it moves its jaws when eating, but was forbidden to be eaten by Mosaic law because it is not cloven-hoofed (Leviticus 11:5).

**Habitat**: Mountains throughout Israel, wherever there are rocky boulders, vegetation and water, holes and crevices among rocks, often on inaccessible hillsides. The Syrian Rock Hyrax is the *Shafan* of the Bible. In Hebrew and Phoenician, it means "the hidden one", referring to its hiding among rocks. It is sometimes confused with the hare or rabbit, which in Hebrew is *Arnevet*.

**Distribution**: Africa and the Middle East. The subspecies *P. c. syriacus* is known from Syria and Lebanon. In Israel, it inhabits Mount Hermon up to 1,300 m., the Golan Heights and Upper Galilee. Allopatric populations are found on Mount Tabor and Mount Carmel. The Sinai subspecies occurs in the Negev Desert and may intergrade with the Syrian subspecies in the Judean Desert. At Ein Gedi, there are two color phases, dark brown and pale grayish-yellow.

## Sinai Rock Hyrax — Figure 8

*Procavia capensis sinaiticus* Ehrenberg, 1828
Shafan Sinai (Hebrew)
Tubsun, Webr or Webeh (Arabic)

**Description**: Head and body 460–495 mm.; ear 25–33 mm.; hind foot 64–77 mm.

The Sinai Rock Hyrax is slightly smaller than the Syrian Rock Hyrax, pale grayish-brown, with narrower upper first molars.

**Habits**: Same as the Syrian Rock Hyrax.

**Habitat**: Similar to the Syrian subspecies, except in the desert, among large rocks and thickets of reeds near oases.

**Distribution**: Mainly an African species. The subspecies *P. capensis sinaiticus* is described from Sinai, (Mount Katarina) up to 2,000 m. In Israel, it is known from the Judean Desert (Ein Gedi) and the Negev Desert.

**Taxonomic Remarks**: The Sinai subspecies was previously regarded as belonging to the Syrian subspecies (Harrison, 1968), but it is morphologically, geographically and ecologically a distinct population.

# Plate 1

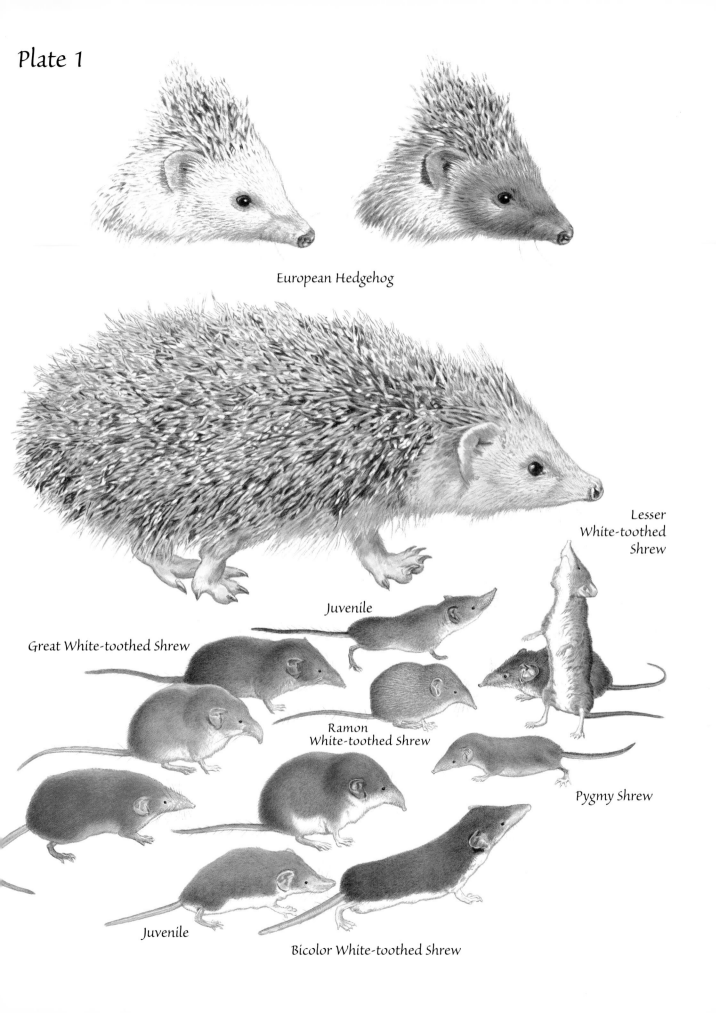

European Hedgehog

Lesser White-toothed Shrew

Juvenile

Great White-toothed Shrew

Ramon White-toothed Shrew

Pygmy Shrew

Juvenile

Bicolor White-toothed Shrew

# Plate 2

Egyptian Long-eared Hedgehog

Ethiopian Hedgehog

Pale Gray Shrew

Thomas's White-toothed Srew

# Plate 3

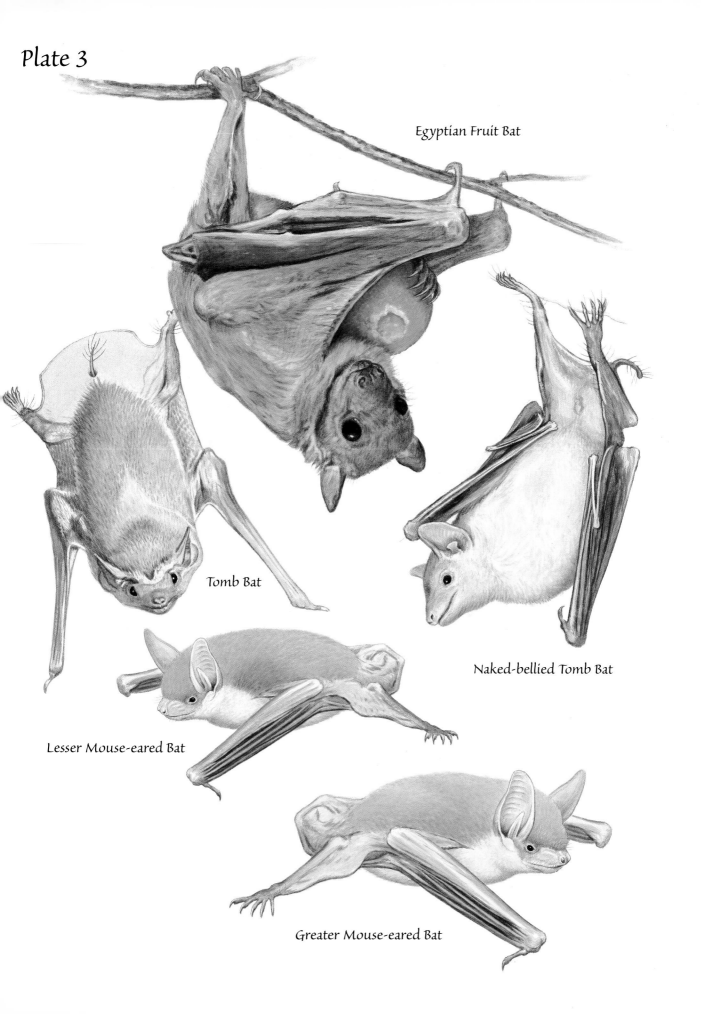

Egyptian Fruit Bat

Tomb Bat

Naked-bellied Tomb Bat

Lesser Mouse-eared Bat

Greater Mouse-eared Bat

# Plate 4

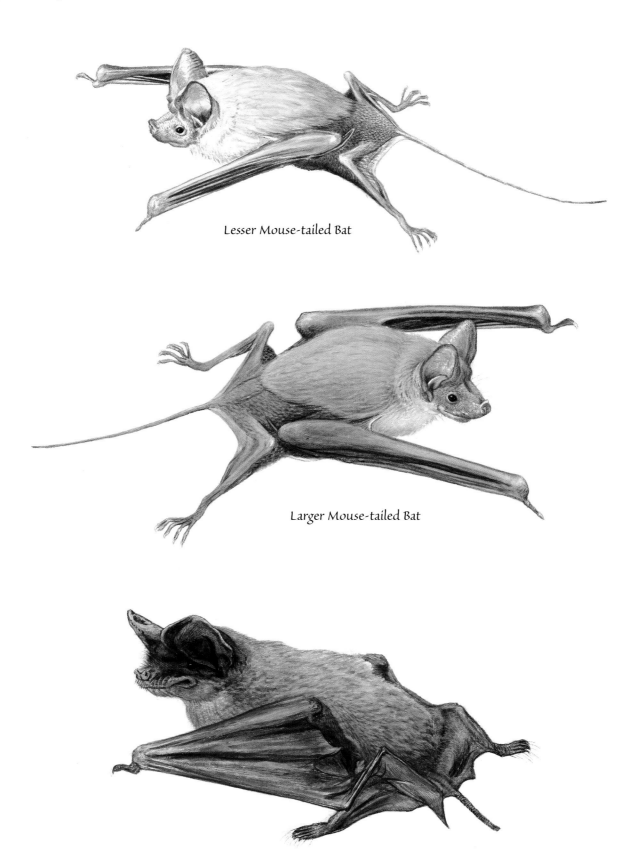

Lesser Mouse-tailed Bat

Larger Mouse-tailed Bat

European Free-tailed Bat

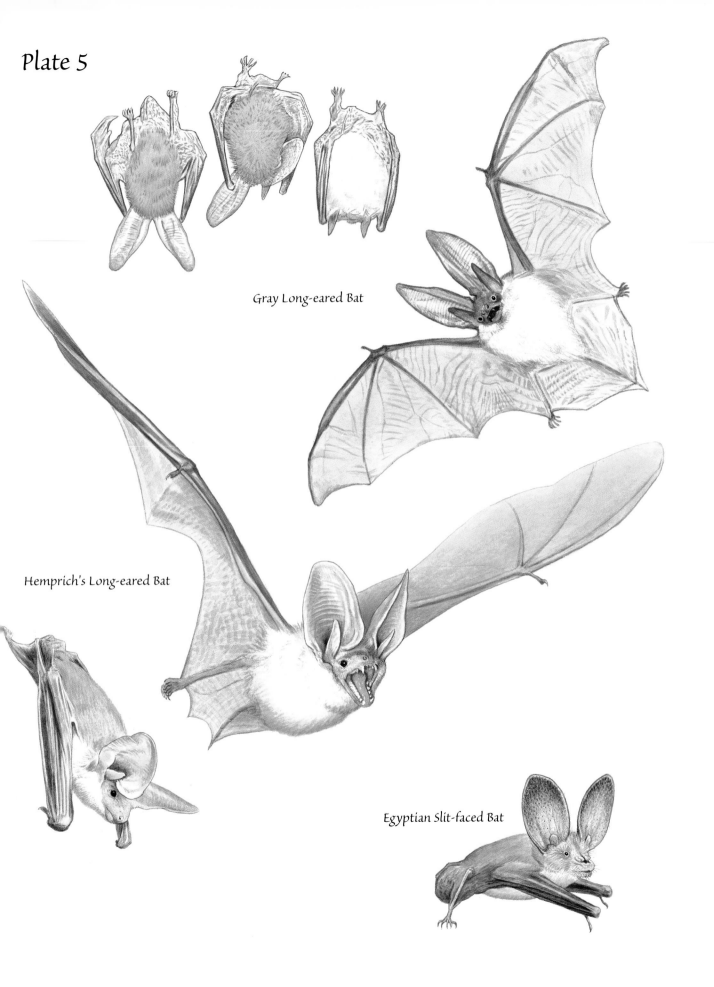

Plate 5

Gray Long-eared Bat

Hemprich's Long-eared Bat

Egyptian Slit-faced Bat

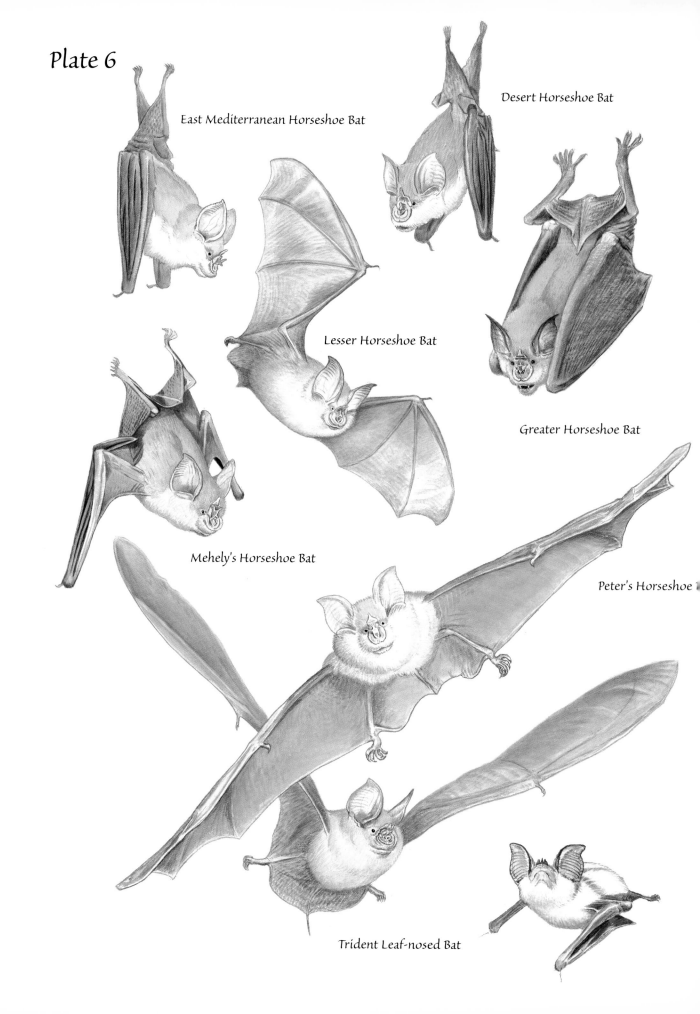

Plate 6

East Mediterranean Horseshoe Bat

Desert Horseshoe Bat

Lesser Horseshoe Bat

Greater Horseshoe Bat

Mehely's Horseshoe Bat

Peter's Horseshoe

Trident Leaf-nosed Bat

Plate 7

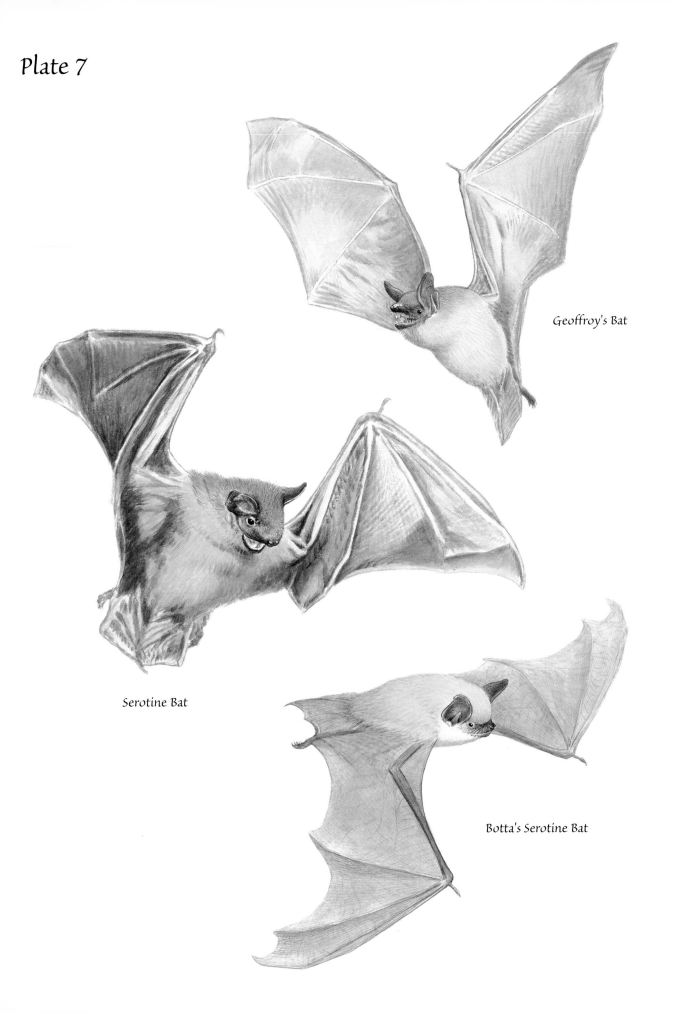

Geoffroy's Bat

Serotine Bat

Botta's Serotine Bat

# Plate 8

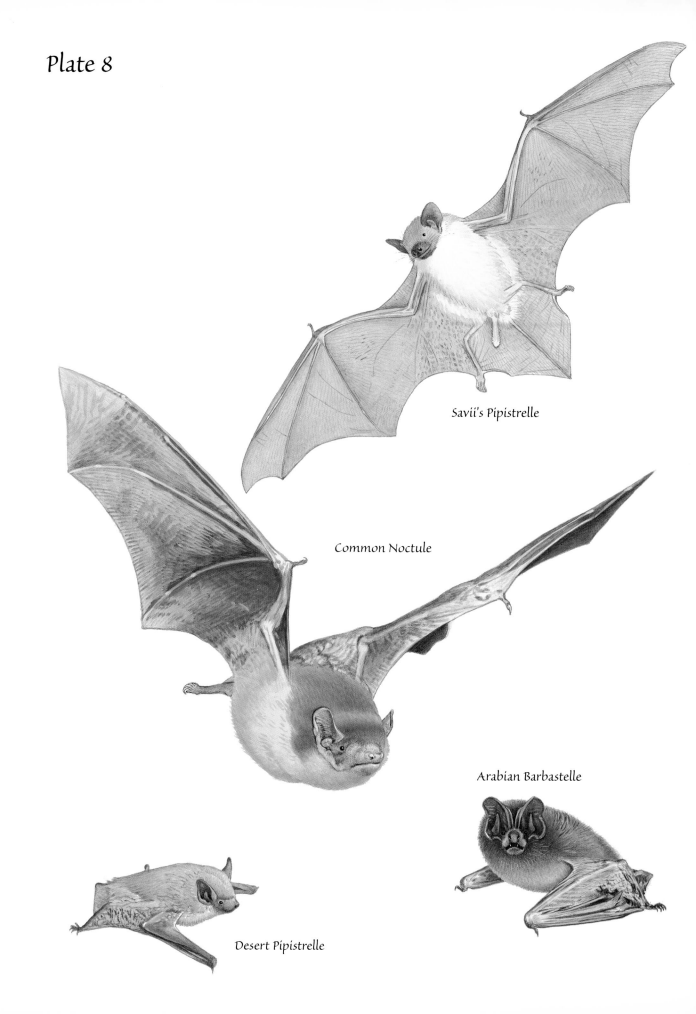

Savii's Pipistrelle

Common Noctule

Arabian Barbastelle

Desert Pipistrelle

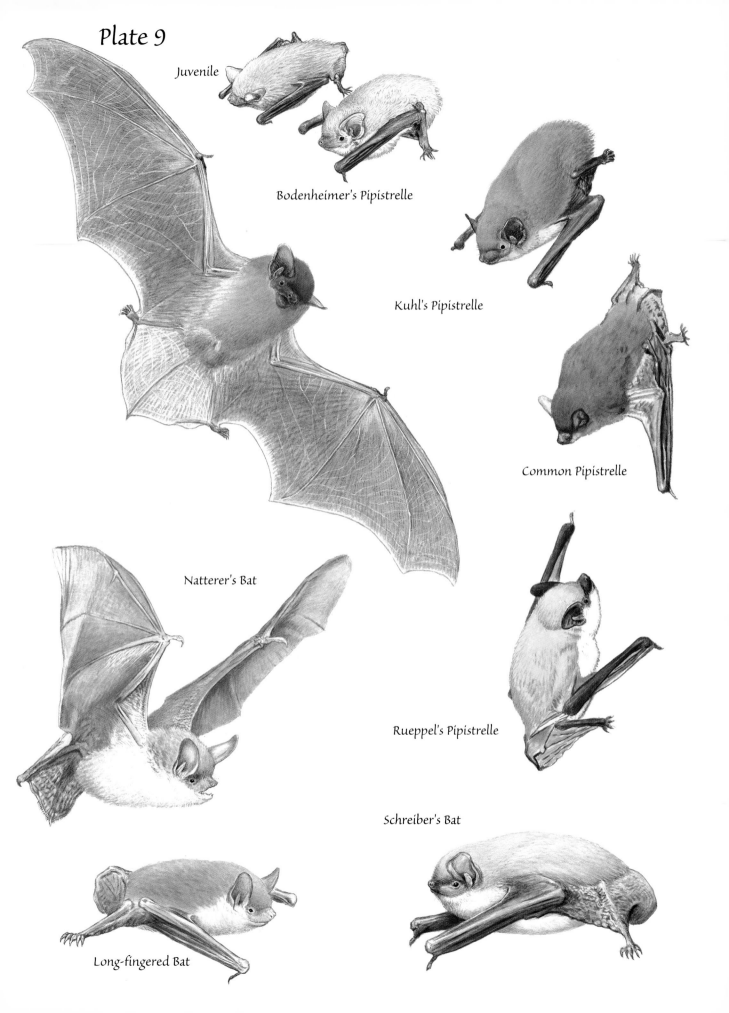

Plate 9

Juvenile

Bodenheimer's Pipistrelle

Kuhl's Pipistrelle

Common Pipistrelle

Natterer's Bat

Rueppel's Pipistrelle

Schreiber's Bat

Long-fingered Bat

Plate 10

Indian Wolf-Dog Hybrid

Arabian Wolf

Plate 11

Indian Wolf

*V. r. rueppeli*

Asiatic Jackal

*V. r. sabaea*

Rueppel's Sand Fox

cub

Common Red Fox

Plate 12

Asiatic Jackal

Arabian Jackal

Plate 13

Arabian Red Fox

Palestine Red Fox

Blanford's Fox

Plate 14

Syrian Brown Bear

# Plate 15

winter

summer

Snow Weasel

Egyptian Mongoose

Common Otter

Plate 16

Syrian Marbled Polecat

Plate 17

Syrian Beech Marten

# Plate 18

Persian Badger

Honey Badger

# Plate 19

Syrian Striped Hyaena

Arabian Striped Hyaena

Plate 20

Palestine Wild Cat

Desert Wild Cat

Plate 21

Sand Cat

Plate 22

Jungle Cat

Plate 23

Arabian Caracal

Plate 24

Syrian Leopard

Plate 25

Arabian Leopard

Plate 26

Above: Persian Lion          Below: Asiatic Cheetah

Plate 31

male

female

Arabian Gazelle

Plate 32

male

female

Dorcas Gazelle hybrid *(isabella phenotype)*

Plate 33

female

male

Dorcas Gazelle hybrid  (dorcas phenotype)

# Plate 34

male
winter

male
summer

female

kid

Sinai Ibex

Plate 35

Syrian Hare

Philistine Hare

# Plate 36

Syrian Hare

Philistine Hare

Sinai Hare

Arabian Hare

Plate 37

Persian Squirrel

Plate 38

Indian Crested Porcupine

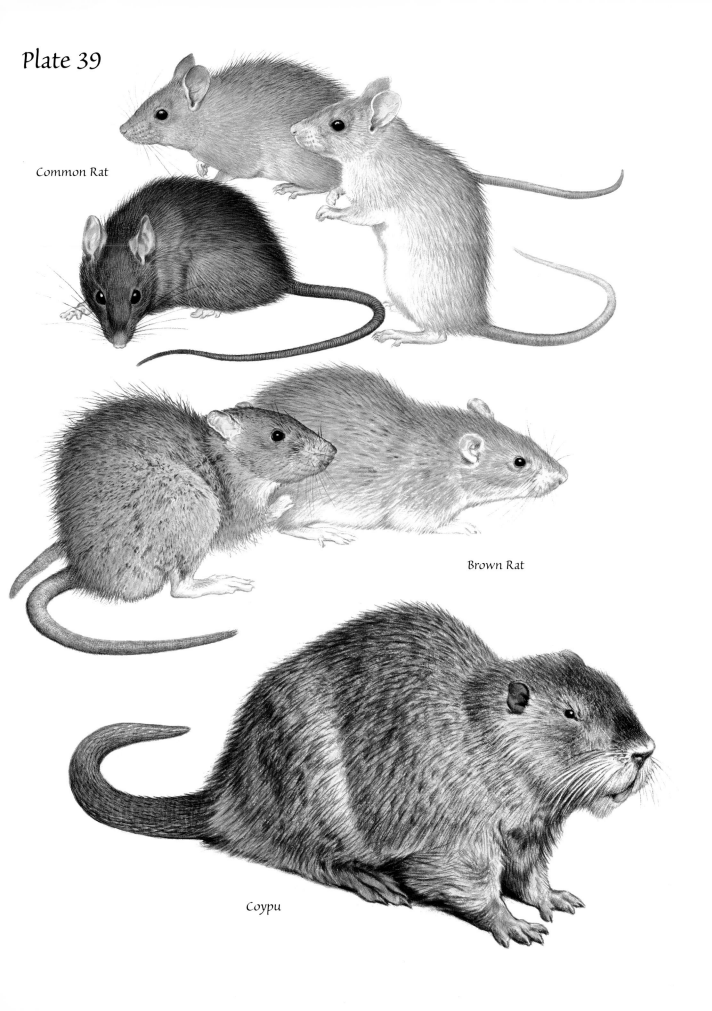

Plate 39

Common Rat

Brown Rat

Coypu

Plate 40

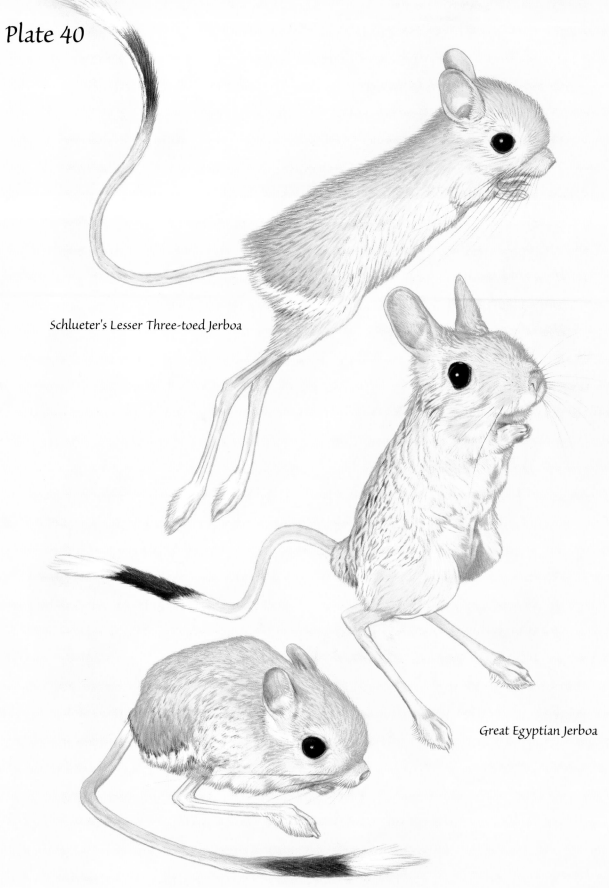

Schlueter's Lesser Three-toed Jerboa

Great Egyptian Jerboa

Wagner's Lesser Three-toed Jerboa

Plate 41

Levant Garden Dormouse

Sooty Garden Dormouse

Plate 42

Forest Dormouse

# Plate 43

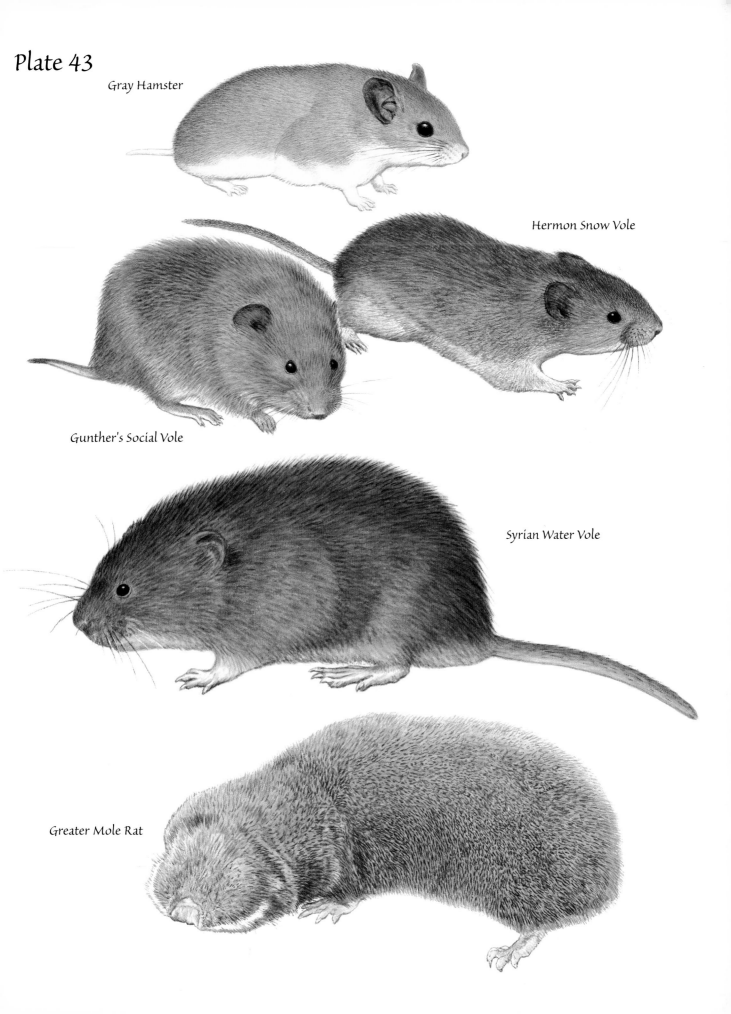

Gray Hamster

Hermon Snow Vole

Gunther's Social Vole

Syrian Water Vole

Greater Mole Rat

Plate 44

Short-tailed Bandicoot Rat

House Mouse

Macedonian House Mouse

juvenile

# Plate 45

Common Field Mouse

Yellow-necked Field Mouse

Alpine Field Mouse

Common Field Mouse

Yellow-necked Field Mouse

Big Levantine Field Mouse

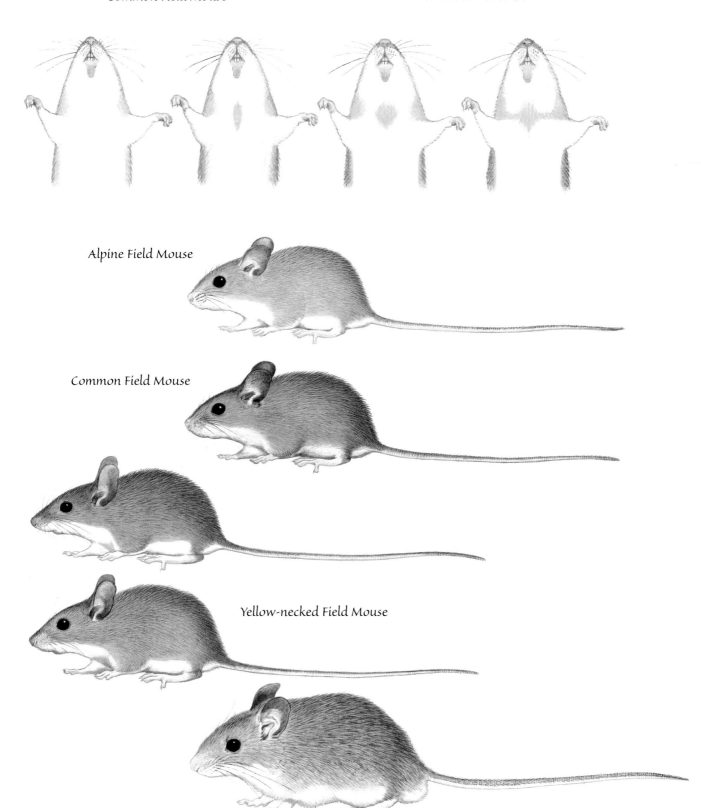

Plate 46

Egyptian Spiny Mouse

Southern Spiny Mouse

Golden Spiny Mouse

Great Egyptian Gerbil

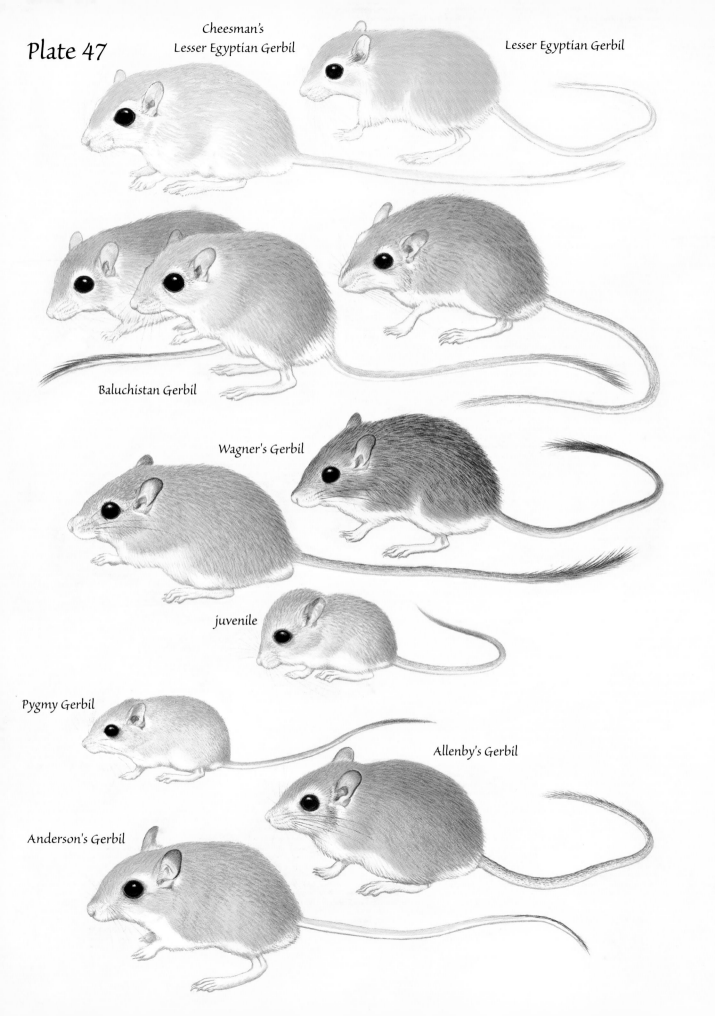

Plate 47

Cheesman's
Lesser Egyptian Gerbil

Lesser Egyptian Gerbil

Baluchistan Gerbil

Wagner's Gerbil

juvenile

Pygmy Gerbil

Allenby's Gerbil

Anderson's Gerbil

# Plate 48

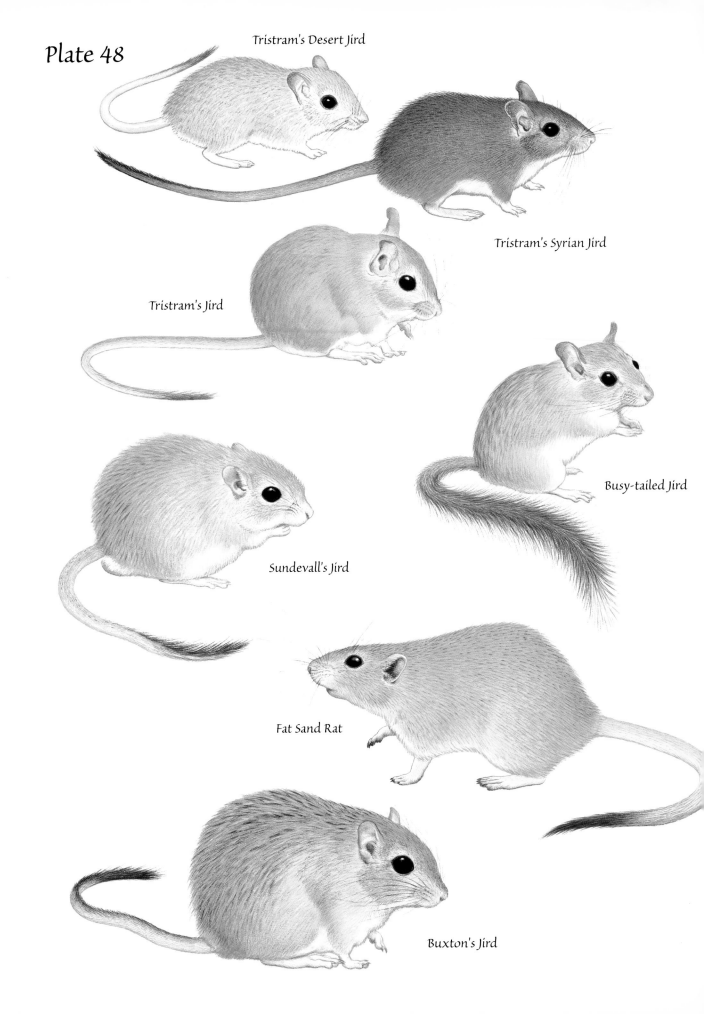

Tristram's Desert Jird

Tristram's Syrian Jird

Tristram's Jird

Busy-tailed Jird

Sundevall's Jird

Fat Sand Rat

Buxton's Jird

# EVEN-TOED UNGULATES: ARTIODACTYLA

Even-toed ungulates are medium- to large-sized herbivorous or omnivorous mammals, characterized by an even number of toes, usually two functional and two nonfunctional, and are commonly called cloven-hoofed animals. They are special for cursorial locomotion. In Israel, they are represented by swine, deer (extirpated in the wild), antelope and ibex.

Fig. 9. Arabian Gazelle, adult male.

# SWINE: SUIDAE

Swine are fairly large; stout and short-legged; and more or less covered with coarse, bristly hair. The snout is rather long, mobile and truncated. The tail is usually small, and the feet each have four toes, but only the center pair are functional. The males have tusk-like canines. They are generally uniformly colored brown and are found in humid as well as arid environments, but near water. Universal in distribution, they are represented in Israel by a single species.

## Palestine Wild Boar                    Plate 28

*Sus scrofa lybicus* Gray, 1868
Hazir Bar (Hebrew)
Hallouf, Khanzir wahshee (Arabic)

**Description**: Head and body 1,240–1,640 mm.; ear 120–170 mm.; hind foot 210–340 mm.; tail 150–200 mm.

The Palestine Wild Boar is a smaller subspecies of the Wild Boar of Europe and Asia. It is the only wild pig found in Israel and is distinguished by its large, heavy-set, laterally compressed body (hog-back). It has a greatly elongated head; large, broad ears; narrow legs; and long, cylindrical tail, which is tufted at the end. A short, thick, upper tusk turns sharply upwards and shears with a lower tusk. The general color is a uniformly-grizzled brown with bristly hairs, shorter in the summer. The snout, ears and legs are blackish. The piglets are spotted and striped lengthwise, but lose their markings by midsummer.

**Habits**: Active mainly at night, it excavates burrows and swims well. The gestation period is 112–175 days and the litter 2–12 in size. Omnivorous, it digs for roots, bulbs and tubers. It also eats fungi and fruit. Contrary to popular belief, it does not overeat. They are gregarious and band together in small numbers. During the day, they wallow in mud and rest in the shade of thickets.

**Habitat**: Forests, tall grass, reed beds, dense thickets and cultivated fields.

**Distribution**: Widespread through Europe, Asia Minor, North Africa and Asia as far east as Japan. The subspecies *S. s. lybicus* occurs in southwestern Asia Minor and Syria. In Israel, it survives in the Golan Heights, Upper Galilee, the Huleh and Jordan Valleys, hills of Samaria and Judea and was extirpated from the coastal plain. A relic population survives along the Jordan River and south of the Dead Sea.

# DEER: CERVIDAE

Deer are fairly large, ruminating ungulates, characterized by two large and two small hoofs on each foot; and by antlers, which are usually borne by the males, and grow larger. They are at first covered with fur-covered skin ("velvet"), and shed each autumn when fully grown. Females seldom have antlers. Their color is generally brown, with or without spots. They are widespread in both the New and Old Worlds, but are found mostly in the northern hemisphere. In Israel, they were formerly represented by four species.

## Mesopotamian Fallow Deer, Persian Fallow Deer

Plate 29

*Dama mesopotamica* Brooke, 1875
Yachmoor (Hebrew)
Ayyal (Arabic)

**Description**: Total length 2,230–2,410 mm.

The Mesopotamian Deer is of medium size. In summer it is brightly-colored reddish-brown, with white spots and white dorsal and lateral stripes. In winter, the coat is more grayish-brown, and the spots are obscure. The tail is all white. The Mesopotamian Deer is distinguished by its antlers, which are much less palmated and spreading, and more vertical than those of the European Fallow Deer *Dama dama*. In the second year, an unbranched spike appears. Each successive year, the new beam grows progressively longer, thicker and forked, with additional tines and snags. In the third year, there is a brow-tine and possibly a trez-tine. The lower beam is flattened, the brow-tine is short and the trez-tine is large and upcurved. The main palmation of the beam is below, rather than above, the middle of the length of the antler. The summit, or crown, is only moderately flattened, breaking up into several snags on the hind border. Palmation width is 7–8.5 cm. There are as many as 12 tines and snags on a single antler. The antlers are acquired in the spring and shed in late winter.

**Habits**: Active in the morning and evening, it travels in small groups. In September or October, during the rutting season, the males establish harems by bellowing and fighting rivals. A single fawn is born after about 230 days. Its diet is mainly grasses.

**Habitat**: Riverine thickets, dense jungles of Tamarisk and Poplar.

**Distribution**: Western Iran, south Asia Minor and Iraq. In Israel, it is known from the Upper Pleistocene of the Huleh Valley (Bnot Ya'acov). It was formerly found in Upper Galilee, Mount Tabor, Mount Carmel and the coastal plain (Emeq Hefer). It has been reintroduced to the country and is thriving in captivity. About 50 deer were set free in Nahal Bezet, western Galilee, in the late 1990s.

**Taxonomic Remarks**: The Mesopotamian Fallow Deer is considered a subspecies of *D. dama* L.,1758 by Mendelssohn and Yom-Tov (1999), but has been shown to be a distinct species (Ferguson, 1985).

---

### Kurdish Roe Deer                    Plate 29

*Capreolus capreolus coxi* Cheesman and Hinton, 1923

Ayal HaCarmel (Hebrew)

Ghazal el Jabal, or Yahmour (Arabic)

**Description**: Head and body 950–1,350 mm.; tail 20–40 mm.

The Kurdish Roe Deer is small and distinguished by the color of its winter coat, a cold, pale gray. The mid-dorsal region, from the nape to the rump, is washed with a darker brown. The underparts are white, and the tail is an inconspicuous stump. The summer coat of this race is unknown. In Europe, it is a bright chestnut-red. The fawn is spotted with white. The antlers appear in the second year as single prongs. By the third year, they are complete, up to 2,300 mm. in height, forked, with a short tine pointing forwards. Those of the fourth year have an additional tine pointing backwards. They are shed in November and renewed by April.

**Habits**: Mainly nocturnal and shy, it inhabits dense thickets. Although not gregarious, it forms family groups. It can swim very well. It is the only deer known to have delayed implantation. After a gestation period of approximately 126 days, twins are born, each at a different location. It feeds on grass.

**Habitat**: Sparsely wooded valleys and the lower slopes of mountains, not higher than 2,400 m.

**Distribution**: Europe and Asia. The subspecies *C. c. coxi* is known from Iraq, Turkey and Lebanon. In Israel, it was common in Upper Galilee, the Jordan Valley and Mount Carmel, where it probably reached the southernmost limit of its range. It has been extirpated since about 1912. There is a report of one caught in 1941 at Ein Sachne (Kibbutz Kfar Szold), which may have strayed from Lebanon. Sightings of deer in recent years refer to the Mountain Gazelle, with which they are often confused. The species has been reintroduced to the country in captivity. In 1996, nine deer were set free on Mount Carmel and appear to be thriving.

# OXEN, ANTELOPE, SHEEP AND GOATS: BOVIDAE

The family of bovines are all ungulates and are characterized by their cloven hoofs and unbranched, hollow horns that are never shed. Both sexes bear horns, although those of the male are larger.

The Wild Goat or Bezoar, *Capra aegagrus*, from Asia Minor, is an ancestor of the domestic goat. In inhabited Mount Carmel in prehistoric times and was introduced from Crete into Israel where several escaped captivity on Mount Carmel.

## OXEN: BOVINAE

Oxen are the largest of the bovines. These include wild and domestic cattle. They are distinguished by their horns, which stand out laterally. Oxen are universal in distribution. It is doubtful whether the Aurochs or Wild Ox, *Bos primigenius*, was present in Israel during historic times. The Indian Water Buffalo, *Bubalus bubalus*, was introduced into Israel in a domesticated state.

## ANTELOPES: ANTELOPINAE

Antelopes are even-toed, ruminating, ungulates, which range greatly in size. They are characterized by two large and two small hoofs on each foot, and by permanent horns with transverse rings, usually borne by both sexes, although those on the male are larger. Graceful and often light in build, they run swiftly and usually associate in herds. They vary in color from blackish-brown to almost white, with every shade of brown in between, and often have stripes or spots. The true antelopes are confined to the Old World, some in Asia but mostly in Africa. In Israel, they are represented by two species. Although the Arabian Oryx, *Oryx leucoryx*, was introduced into the Arava Valley, its previous occurrence in Israel (Western Palestine) has not been substantiated.

101

## Mountain Gazelle, Chinkara    Plate 30

*Gazella gazella gazella* Pallas, 1766

Tsvi Israel (Hebrew)

Idmi or Edmi (Arabic in North Africa, Iraq and Arabia)

Ghasel (Arabic in Western and Southern Arabia and Israel)

**Description**: Head and body 910–1,150 mm.; ear 110–125 mm.; hind foot 320–354 mm.; tail 80–130 mm.

The Mountain Gazelle is of medium size. It is distinguished by the dark, smoky-fawn upper parts in the winter coat; a distinctly darker flank band; and a rufous-fawn central face band with a distinct, blackish spot at the top of the nose. The white of the interior upper foreleg does not extend to the outer side, as in the Dorcas Gazelle. In the summer, the color is a paler fawn, and the flank band is less conspicuous. The horns of the male are 150–291 mm. long; wide apart and divergent; not strongly lyrate; with a sigmoid curve in profile; and the tips variably hooked forwards and slightly inwards. The horns of the female are very short, 85–115 mm., slender and variably upturned at the tips. The tail is black.

**Habits**: Gregarious, they travel in groups of 5–10 and are active in the morning and evening. Gestation is about 180 days. The litter is one, rarely two. Males engage in sparring with their horns during the mating season. They feed on plant material. When excited, the white rump expands.

**Habitat**: Mountains, plains, valleys and cultivated areas.

**Distribution**: North Africa, Iran, Syria and Arabia. The nominate subspecies *G. g. gazella* occurs in Syria and Lebanon. In Israel, it is fairly common in the northern half of the country, south to the Judean hills.

## Arabian Gazelle    Plate 31, Figure 9

*Gazella gazella cora* Smith 1827

Tsvi HaArava (Hebrew)

Idmi (Arabic)

**Description**: Head and body 1,000–1,162 mm.; ear 130.5–140.1 mm.; tail 155 mm.

The Arabian Gazelle, *G. gazella cora*, is distinguished by its slender build; large ears; and broad black nose; spot, black flanks and pygal stripe. The smoky-pink or fawn upper flanks of the male may be flecked with some small whitish spots. The white of the upper inner foreleg does not extend to the outer foreleg and thus is not visible in profile, as in the Dorcas Gazelle. The horns of the male are divergent and turn inwards at the tip. They are 208–277 mm. in length. The female's horns are shorter and straight.

**Habits**: It is active in the morning and evening. If disturbed, it retreats to dense vegetation. A single fawn is born after a gestation period of 180 days. It feeds on Acacia, often standing on its hind legs only. It is not dependent on free water, due to a diet of largely succulent plants.

**Habitat**: Low hills and wadis in deserts with dense Acacia stands and shrubby thickets.

**Distribution**: North Africa, Iran, Syria and Lebanon. The subspecies *G. g. cora* was formerly found in Israel in the northern Arava Valley (Hatseva), but is now known only from the southern Arava Valley (Yotvata).

**Taxonomic Remarks**: The desert subspecies of *G. gazella* in Israel was described as *G.g. acaciae* Mendelsohn, Groves and Shalmon, 1997.

The sample of five specimens for the desert population of *G. gazella* in Israel is statistically insufficient to recognize it as a new subspecies. Some comparative measurements for *G. gazella cora* are lacking or limited. Morphologically *G. gazella acaciae* is very similar to *G. gazella cora* and may represent only a new variant of a highly variable subspecies.

---

## Dorcas Gazelle      Plates 32, 33, Figure 11

*Gazella dorcas* Linnaeus, 1758
Tsvi HaNegev or Tsvi HaMidbar (Hebrew)
Afri (Arabic in Arabia)

**Description**: Head and body 815–1,010 mm.; ear 130–150 mm.; hind foot 270–337 mm.; tail 104–145 mm.

The Dorcas Gazelle may be distinguished by its small size. It has distinct facial markings, with a strong rufous central facial band, gray nose and large, gray ears. An indistinct rufous-brown lateral band sharply demarcates the variable fawn upper parts from the white underparts. There is no dark pygal band, and the rump has a white patch. Unlike the Mountain or Arabian Gazelle, the white on the inner side of the upper foreleg extends to the anterior outer side of the leg. The fawn-colored forelegs have long, rufous or blackish knee "brushes". The horns of the male are 175.5–236 mm., more or less lyrate-shaped, and their tips incline towards each other. The female has shorter horns, 123–199 mm. The Dorcas Gazelle in Israel, *G. d. dorcas* X *G. d. isabella*, shows great variation which represents a hybrid population between the North African race *G. d. dorcas* and the Red Sea subspecies *G. d. isabella* (Ferguson, 1981b). The following are diagnoses of each race:

*Gazella dorcas dorcas* L, 1758

Small, pale, sandy-fawn upper parts. Lacks black nose spot and has reddish-fawn knee brushes. Male horns heavy, strongly lyrate, widely divergent from base, strongly compressed laterally, slightly convergent towards tips, which curve inwards and upwards. Widest space between horns near middle. Female horns much smaller, more slender, not lyrate, curved backwards, rather goat-like, fully ridged, tips not strongly convergent.

*Gazella dorcas isabella* Gray, 1846 (= *G. littoralis* Blaine, 1913)

Larger and browner or deeper reddish color, less sandy in tinge. Distinct black nose spot and black knee brushes. Male horns light, narrowly divergent from base, slightly compressed laterally, nearly straight, strongly convergent towards tips, which are hooked inwards and forwards as much as a right angle. Widest space between horns below tips. Female horns similar to male only smaller, more slender, lyrate, ridged only near the base, upright tips strongly convergent.

**Habits**: Active during the morning and evening, it associates in groups of 3–7, sometimes as many as 20 or more. Gestation lasts 169–181 days. The one or two young are able to stand almost immediately after birth and can run in a week. It feeds on plants and Acacia.

**Habitat**: Mountains, steppes and plains in deserts and semi-deserts, often in rocky wadis with Acacia trees and vegetation.

**Distribution**: North and East Africa, Sinai and Arabia. In Israel, it inhabits the western side of the Dead Sea and the southern half of the country.

**Taxonomic Remarks**: Mendelssohn (1974) remarks that the Israeli Dorcas Gazelle is similar to *isabella*. Groves (1983) suggests that

it is possible that all the Dorcas Gazelles from the Sinai Desert and Israel may be referred to *G. d. isabella*. Qumsiyeh (1996) states that the differences between the subspecies *G. d. dorcas* and *G. d. isabella* are minimal and clinal in nature. The fact is the two subspecies are so distinct that *G. d. isabella* is sometimes considered a distinct species (Dorst and Dandelot, 1970). Mendelssohn and Yom-Tov (1999) now state that the subspecies of the Dorcas Gazelle in Israel is *G. d. littoralis* (Blaine, 1913). The subspecies *littoralis* is a junior synonym of *G. d. isabella* (Millier, 1912; Groves, 1981). Qumsiyeh also claims that many of the subspecies identifications are based on few specimens, without genetic studies or proper statistical analysis of morphological differences. The Dorcas Gazelle population in Israel does not indicate a character gradient between subspecies, as speculated by Qumsiyeh, but a zone of secondary hybridization, based on 34 specimens showing a mixture of traits between *G. d. dorcas* and *G. d. isabella*, with some hybrid parental phenotypes characteristic of each subspecies (Ferguson, 1981b) (Fig. 10).

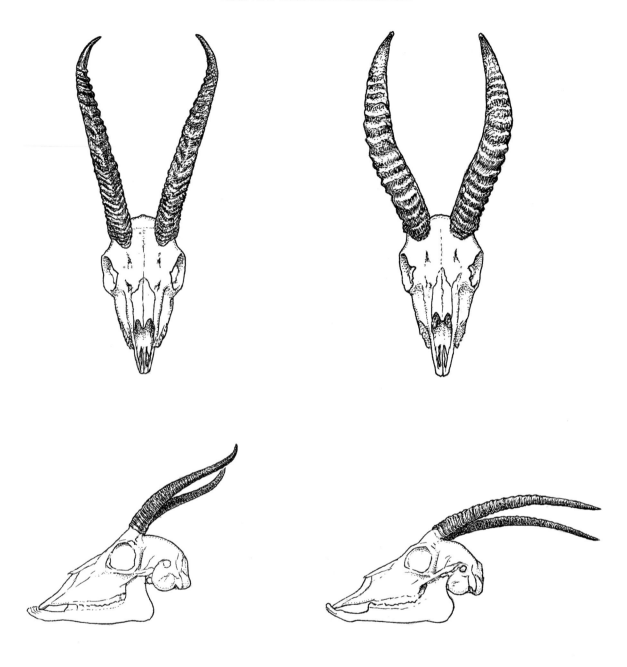

Fig. 10. Parental phenotypes of the hybrid Dorcas Gazelle in Israel and Sinai. Upper left: Male *G. d. isabella* type, M6558 Mitla, Sinai. Upper right: Male *G. d. dorcas* type, M4676, Ein Geddi, Israel. Lower left: Female *G. d. isabella* type, M4446, Sde Boker, Israel. Lower right, Female *G. d. dorcas* type, M3794, El Arish, Sinai. All specimens in the collection of the Zoological Museum of the Tel Aviv University.

## SHEEP AND GOATS: CAPRINAE

Intermediate species of sheep and goats are so closely related that the genus of one seems to pass gradually into the other. Typical sheep are characterized by their massive horns, curved in a spiral down the sides of the head. The horns of goats are more erect and not as compactly curved. Sheep have "tear glands" below their eyes which goats do not have. Sheep also have glands on their hind feet, while goats do not and sometimes not on their front feet either. In addition, the male goat usually has a beard. Generally, wild sheep and goats have crisp hair and short tails. They vary in color from dark brown to white, sometimes with a pattern of markings. Sheep and goats are found mostly in the northern hemisphere. In Israel, there are no wild sheep and only one species of wild goat, the ibex.

### Sinai Ibex                    Plate 34, Figure 11

*Capra ibex sinaitica* Hemprich and Ehrenberg, 1828
Ya'el (Hebrew)
Beden (Arabic)

**Description:** Head and body 1,022–1,367 mm.; ear 100–120 mm.; hind foot 250–320 mm.; tail 56–83 mm.

The Sinai Ibex is the only wild goat found in Israel. It is of medium size, more lightly built than the ibex of Eurasia, with the male much larger than the female. The summer coat is a yellowish-fawn, becoming darker grayish-brown in the male in winter, when the black markings on the stocky legs increase in area. There is a blackish stripe along the back, dark across the shoulders and chest, with a black beard and tail. The ears are rather large, with three dark streaks on the inner side. The male is distinctive for his scimitar-shaped horns, 425–580 mm. in length on the outside curve, with bold transverse knobs on the front surface.

One to three ridges are added to the horn every year, but the annual rings are more reliable indictors of age. The female's horns are much smaller, more erect and smoother, 165–215 mm. in length. The juvenile resembles the female.

**Habits:** Gregarious, they usually associate in small groups, with a single, old male leading his harem and young during the rutting season. Since the forelegs are somewhat shorter than the hind legs, it seeks safety by running uphill. Rival males fight by rising on their hind legs and with heads tilted, crashing their horns together. Outside the rutting season, males and females remain apart. One or two kids are born after a gestation period of 147–180 days. Ibex graze and browse on vegetation, sometimes standing on its hind legs only.

**Habitat:** Dry desert mountains, deep wadis and steep cliffs, not far from water on which they depend.

**Distribution**: Europe, Asia, southwest Asia, Egypt, Sudan, Eritrea and Arabia. In Israel, the subspecies *C. i. sinaitica* is confined to the Dead Sea region and Negev Desert, south to Eilat.

**Taxonomic Remarks**: The ibex of Israel and the Sinai desert have been regarded as the subspecies *C. ibex nubiana*, F. Cuvier, 1825. The population of Israel and the Sinai desert, however, have been geographically and ecologically isolated from the African Nubian Ibex for the past several million years by the Isthmus of Suez. Morphologically, *C. ibex sinaitica* seems to differ in having less black on its legs, regardless of age or winter pelage (Fig. 11).

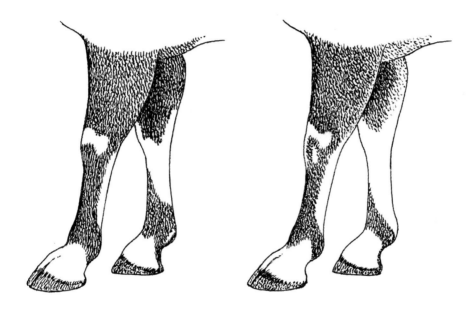

Fig 11. Patterns of the forelegs of ibex.
Left: Nubian Ibex, *C. i. nubiana*. Right: Sinai Ibex, *C. i. sinatica*.

# HARES AND RABBITS: LAGOMORPHA

Hares and rabbits were previously classified as rodents (Order Rodentia) because of their gnawing incisors. They differ from rodents, howewer, in possessing a second pair of small, peg-like incisors behind the long central pair in the upper jaw and in seriological studies.

Fig. 12. Leveret of Syrian Hare

# HARES: LEPORIDAE

The family of hares is represented by a single genus, and they are all quite similar in appearance. Females are usually larger than males. Moderately large for rodents, they are characterized by their long ears and long hind legs. The tail is short, upturned and cotton-like. Each forefoot has five toes, while each hind foot has only four toes. The toes and soles of the feet are covered with brush-like hairs. They are generally brown in color with a mixture of gray or reddish-brown or white. Strictly terrestrial and cursorial, they inhabit open lands. Hares are cosmopolitan and are represented in Israel by a cline of a single species. There are no rabbits in Israel.

## Syrian Hare,                                        Plates 35, 36, Figure 12

*Lepus capensis syriacus* Ehrenberg, 1833

Arnevet Surit (Hebrew)

Arnel (Arabic)

**Description:** Head and body 422–653 mm.; ear 101–126 mm.; hind foot 108–145 mm.; tail 63–88 mm.

The Syrian Hare is a race of the Common Hare of Europe, northern Asia and Africa. It may be distinguished by its fairly large size; thick fur; long ears; grizzled, brownish-gray upper parts; and light gray sides and thighs. The breast and forelimbs are tinged with rufous, and there is a distinct rufous patch on the nape. The belly is white, and the upper side of the tail and the hind tips of the ears are black.

**Habits:** Mostly nocturnal, it hides during the day in its "form", or hiding place, in a slight depression among the grass or beneath a bush. Here, 2–4 leverets, sometimes more, are born after a gestation of 42 days. Hares are extremely rapid runners. During the rutting season, the females rebuffs the male in stand-up boxing matches, leaping over each other and kicking with their hind feet. They feed on plants and are destructive to agriculture.

**Habitat:** Steppes, open country, cultivated fields and coastal sand dunes.

**Distribution:** Widespread across Europe, northern Asia, Asia Minor, Arabia, and Africa. The subspecies *L. c. syriacus* is known from Lebanon, Syria and northern Jordan. In Israel, it is common in the northern half of the country, south almost to Kiryat Gat. Part of a cline, it may intergrade with other races to the south.

## Philistine Hare                                        Plates 35, 36

*Lepus capensis philistinus* ssp. nov.

Arnevet Philistini (Hebrew)

Arnel (Arabic)

**Description:** Head and body 450–505 mm.; ear 110–125 mm.; hind foot 118–138 mm.; tail 58–80 mm.

109

The Philistine Hare is a substrate subspecies of the Common Hare, *Lepus capensis*. It appears to be slightly smaller than the Syrian Hare and distinctive for its uniform, reddish-fawn coloration and lack of black, except for the tips of the ears and the tail. It is larger than the Sinai subspecies, with relatively shorter ears.

**Habits**: Presumably similar to other subspecies of the Common Hare.

**Habitat**: Open country and cultivated fields.

**Distribution**: The species *Lepus capensis* is very widespread, polymorphous and clinal, ranging across Europe, Asia Minor, Arabia, northern Asia and Africa as far south as South Africa. The subspecies *L. c. philistinus* is found in the southern coastal plain between Qedma and Beersheba. It may intergrade with *L. c. syriacus* to the north and *L. c. sinaiticus* to the south.

**Taxonomic Remarks**: The Philistine Hare seems to have adapted to a substrate of red Hamra soil, but wind-blown gray loess soil has covered much of the region where it has lived over the past several thousand years, not long enough for it to have adapted in color to the grayer topsoil or possibly due to plowing.

**Diagnosis**: Since the population of *Lepus capensis* in the southwest coastal plain of Israel can be distinguished from the dark Syrian race to the north, and the pale grayish Sinai race to the south, by its reddish-fawn color without any black, it is suggested that it represents a distinct geographical population and be known as

*Lepus capensis philistinus* ssp. nov.

**Holotype**: TAU M3293. Deposited in the Zoological Museum of Tel Aviv University.

**Type locality**: Qedma

**Collector.**: M. Yaniv

**Etymology**: Named after the region in which it is found.

---

## Sinai Hare                                  Plate 36

*Lepus capensis sinaiticus* Ehrenberg, 1833
Arnevet Sinai (Hebrew)
Arnel (Arabic)

**Description**: Head and body 400–459 mm.; ear 103.5–129 mm.; hind foot 93–109 mm.; tail 65–74 mm.

The Sinai Hare is a desert subspecies of the Common Hare of medium size, smaller than the races to the north, with relatively large ears and pale sandy upper parts. It has reduced black ear tips, and there is a whitish post orbital stripe. The nape is pale rufous.

**Habits**: Presumably similar to other races of the species.

**Habitat**: Stony desert plains and wadis with scrubby vegetation.

**Distribution**: Widespread across Europe, Asia Minor, northern Asia and Africa. The subspecies *L. c. sinaiticus* is known from eastern Sinai and the southern half of Israel, generally south of the 100 isohyte line. It may intergrade with the Philistine Hare to the north.

## Arabian Hare                    Plate 36

*Lepus capensis arabicus* Ehrenberg, 1833
Arnevet Aravi (Hebrew)
Arnel (Arabic)

**Description**: Head and body 291–410 mm.; ear 92–127 mm.; hind foot 73–100 mm.; tail 30–90 mm.

The Arabian Hare is a desert subspecies of the Common Hare, distinguished by its small size; relatively large ears, with a black tip; and shorter hind legs. It is generally darker than the Sinai Hare, with grayish-brown upper parts, variably grizzled with black.

**Habits**: Presumably similar to other subspecies of the species.

**Habitat**: Arid deserts.

**Distribution**: Iraq, Kuwait, Saudi Arabia and Yemen. In Israel, it is rare and confined to the southern Arava Valley (Bir Hindis). Yom-Tov (1967) listed *L. c. arabicus* from Eilat.

**Taxonomic Remarks**: Atallah (1977) recognized only two subspecies in the eastern Mediterranean region, *L. c. syriacus* and *L. c. arabicus*, and reassigned *L. c. sinaiticus* and *L. c. judeae* Gray, 1876 in Israel as synonyms of *L. c. arabicus*. Harrison and Bates (1991) retain *L.c. sinaiticus* for southern Israel.

111

# RODENTS OR GNAWING MAMMALS: RODENTIA

Rodents are small- to medium-sized mammals. They are characterized by the fact that they have only one pair of chisel-like teeth above and below, with the exception of the hare which has a second smaller pair behind the front pair. Between these gnawing teeth and the cheek or grinding teeth, there is a pronounced gap. Rodents are generally brownish in color, sometimes blackish or grayish, and rarely with markings. They are usually terrestrial, although some are arboreal or amphibious. The order of Rodents is the largest in the world and is well represented in Israel by six families and 34 species, one of which, the Coypu, was introduced and is now feral.

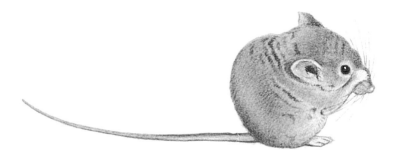

Fig. 13. House Mouse, grooming.

# SQUIRRELS: SCIURIDAE

Squirrels are medium-sized rodents, characterized by their hairy or bushy tails. They have four toes on their forefeet and five on their hind feet. Some have large cheek pouches and hairy soles. They are arboreal or terrestrial and fossorial. Cosmopolitan, they mostly inhabit the temperate and tropical regions of the world. In Israel, they are represented by a single species.

## Persian Squirrel                     Plate 37

*Sciurus anomalus syriacus* Ehrenberg, 1828
Sana'i Zahov (Hebrew)
Sinjab, Jerd el Jabal (Arabic)

**Description**: Head and body 208–213 mm.; ear 23–31 mm.; hind foot 45.8–60 mm.; tail 140 mm.

The Persian Squirrel is a subspecies of the Common Red Squirrel of Eurasia. It is distinguished by its ash-gray upper parts; rufous head and shoulders; pale orange belly; and grizzled, bushy tail. The fairly large ears are surmounted with tufts of hair, longer in the winter.

**Habits**: Mainly diurnal and arboreal, it often descends to the ground. Sometimes, it suns itself on a rock. It builds its nest in a tree. After a gestation period of 38 or 39 days, 5–7 young are born. Its diet consists of nuts, seeds, fruit, pine shoots, birds' eggs and nestlings.

**Habitat**: Wooded areas such as open oak, cedar or pine forests, among olive trees and rocky outcrops at elevations up to 1,830 m.

**Distribution**: Europe, Asia Minor and Asia. The subspecies *S. a. syriacus* is known from Syria, Lebanon and Jordan. It was found in northern Israel at the beginning of the century and was thought to be extirpated, but in 1967, it was rediscovered in Wadi Assal, several kilometers from Dan at the foot of Mount Hermon. It is also known from the Golan Heights.

113

# PORCUPINES: HYSTRICIDAE

Porcupines are fairly large rodents, characterized by stiff, sharp, erectile quills mixed with their hair. The open-ended, hollow spines are generally banded with black and white, whereas the hair may be brownish. They are widespread in distribution, and found in both the New World, where they are arboreal, and in the Old World, where they are strictly terrestrial with five-toed, plantigrade feet. In Israel, they are represented by a single species.

## Indian Crested Porcupine
Plate 38

*Hystrix indica indica* Kerr, 1792

Darban (Hebrew)

Daalej, Nee iss or Nis, Neer, Hirr (Arabic)

**Description**: Head and body 835–885 mm.; ear 40–44 mm.; hind foot 87–100 mm.; tail 105–175 mm.

The typical race of the Indian Crested Porcupine is the largest rodent found in Israel. It is characterized by the sharp, hollow quills that cover most of its body. These consist of short, stout quills, mostly concealed by long, slender ones, and marked with broad bands of black and white. Unlike the Common Porcupine, *Hsytrix cristata*, of southern Europe and Africa, it has a hairy nose; quills at the base of the tail; and sometimes a row of mostly white quills in the middle of the hind part of the back.

**Habits**: It is nocturnal. Strictly terrestrial, it lumbers along with its quills rattling. It is gregarious and burrows in colonies. After a gestation period of about 112 days, 1–4 young are born. It is not aggressive and dangerous only when molested or wounded. Then, it faces the enemy with its back; erects its spines; and rattles them. For protection, the shorter quills are barely held by the skin and come loose on contact, causing the victim painful inflammation. Its diet consists of succulent vegetation, bark, roots, tubers and sometimes carrion.

**Habitat**: Hills, valleys and plains.

**Distribution**: India, Iran, Iraq, Syria and southern Arabia. In Israel, it is fairly common throughout the country.

# COYPUS: CAPROMYIDIAE

The Coypu or Nutria is a large rodent, with partially webbed hind feet and a rather bare cylindrical tail. It is amphibious and inhabits marshes, ponds and rivers. A native of South America, it was introduced into Israel where it is now feral.

## Coypu or Nutria                    Plate 39

*Myocastor coypus* Molina, 1782
Nutria (Hebrew)

**Description**: Head and body 430–635 mm.; hind foot 125–140 mm.; tail 255–445 mm.

The Coypu may be distinguished by its large size. It has coarse yellowish-brown or reddish-brown guard hairs, which nearly obscure the gray underfur. The tip of the muzzle is white; and it has short, rounded ears; webbed hind feet; and a long, sparsely-haired tail.

**Habits**: Amphibious, it swims with the top of its head and back out of the water. It burrows in banks near the water. It has a peculiar, moaning cry during the mating season. After gestation of 120–150 days, 3–9 young are born. It is vegetarian but also eats mollusks.

**Habitat**: Found only near fresh water, around swamps, marshes, fish ponds and rivers.

**Distribution**: Central and southern South America. In Israel, they were introduced for the purpose of fur farming, but some escaped or were set free and are now feral in the Huleh and Beit Shean Valleys, coastal plain (Ma'agan Michael, Ma'ayan Zvi, Alexander River) and in the Negev Desert (Ein Yahav).

## JERBOAS: DIPODINAE

Jerboas are remarkable for their extreme adaptations for saltation. They are characterized by their very long, specialized hind legs, which have only three functional toes; and by their tiny forefeet, which give them a kangaroo-like appearance. Their bodies are compact, with large heads and large eyes. The profile of the head is distinctly concave, due to a high forehead and a thick fold of skin that is drawn over the nostrils when the snout is used for burrowing. A brush of stiff hairs on the ventral surfaces of the phalanges gives additional traction on sand. Locomotion is mainly bipedal, the forefeet used for moving slowly. They are capable of enormous leaps using the long, feather-like tufted tail for balance. When at rest, the tail may be used for support, kangaroo fashion. They are a pale sandy color. Jerboas are especially adapted to inhabit arid or semi-arid regions, in soil that they can burrow into, and are widespread in Asia Minor and North Africa. In Israel, they are represented by two species.

## Greater Egyptian Jerboa     Plate 40

*Jaculus orientalis orientalis* Erxleben, 1777
Yarboa Gadol (Hebrew)
Jarboua, Djerba or Gaurti (Arabic)

**Description**: Head and body 134–180 mm.; ear 27–35 mm.; hind foot 70–80 mm.; tail 226–250 mm.

The typical Greater Egyptian Jerboa of North Africa is distinguished by its large size. It has a long hind foot and sandy-fawn upper parts, more or less variegated with black. There is a white stripe across the thigh, and blackish-brown hairs along the under surface of the hind feet and toes. The underparts are white. The distal third of the tail is encircled with black and has a white tip.

**Habits**: Nocturnal and colonial, it burrows in the ground. The entrance is plugged against the heat during the day. The gestation is 42 days, and the litter varies from 2–6. It moves about on all fours and leaps with its hind legs as much as 1.5–3 meters, in defense, but usually walks bipedally. It feeds on succulent plants and seeds and does not require free drinking water.

**Habitat**: Open deserts and fallow fields.

**Distribution**: North Africa, Algeria, Tunis, Libya and Egypt. In Israel, it has been recorded from the northern Negev Desert (northeast of Beersheba) and western Judean Desert (Arad).

## Thomas's Lesser Three-toed Jerboa, Plate 40
## Lesser Egyptian Jerboa

*Jaculus jaculus vocator* Thomas, 1921
Yarboa, Matsui (Hebrew)
Jarboa, Djerba or Gaurti (Arabic)

**Description**: Head and body 92–120 mm.; ear 19–26 mm.; hind foot 51–64 mm.; tail 128–203 mm.

Thomas's Lesser Egyptian Jerboa, *J. j. vocator*, is an eastern subspecies of the North African *Jaculus jaculus*. Although there is an overlap in size, *J. j. vocator* is generally smaller than the subspecies *J. j. macrotarsus* or *J. j. shlueteri*, and usually has a smaller hind foot. Mendelssohn and Yom-Tov (1999) have the means of length and tail of *J. j. vocator* larger then those of *J. j. shlueteri*, and the female reaches the size of the Greater Egyptian Jerboa! Those inhabiting a sandy desert or dunes are more buffy, while those living in a stony desert appear more grayish. Color does not seem to distinguish subspecies of this species. A black band completely encircles the tail near the end.

**Habits**: It is nocturnal and crepuscular. See Schlueter's Jerboa.

**Habitat**: Arid, sandy or stony desert plains with some vegetation.

**Distribution**: Southwestern Iran, Arabian Peninsula, North Africa and the Sinai desert.

The subspecies *J. j. vocator* occurs in southeastern Syria, eastern Jordan, Saudi Arabia and Kuwait. In Israel, it occurs in the Arava Valley and may penetrate the southern Jordan Valley.

**Taxonomic Remarks**: In Israel, there is no ecological connection between psammophiles (sand-adapted species) of the west with those of the east, except maybe around the perimeter of Sinai, so that the population of Jerboas in the northwestern Negev Desert is essentially isolated from that in the Arava Valley. It is doubtful, therefore, that the race *J. j. vocator* occurs in the northwestern Negev Desert, but may very well be represented in the Arava Valley (Bir Melikha).

The small, nominate subspecies, *J. j. jaculus* of Egypt and North Africa does not occur in Israel, as was once believed (Bodenheimer, 1935), but reaches the Sinai Desert east of the Suez Canal.

## Wagner's Lesser Three-toed Jerboa    Plate 40

*Jaculus jaculus macrotarsus* n. status
Yarboa Matsui (Hebrew)
Jarboa, Djerba or Gaurti (Arabic)

**Description**: Head and body 110–128 mm.; ear 20–23 mm.; hind foot 64–78 mm.; tail 165–185 mm.

Wagner's Lesser Three-toed Jerboa, *Jaculus j. macrotarsus*, is similar to Schlueter's Jerboa in that it has a very large hind foot. However, *J. j. macrotarsus* can easily be distinguished from the latter by the ventral aspect of the tail. The black band near the end of the tail is typically solid

black, unlike Schlueter's, where the black band is interrupted on the underside by a white streak down the middle. It can also be distinguished by its grayish ears; black and white vibrissae; and dusky, hairy soles of the hind feet.

**Habits**: Similar to Schlueter's Jerboa.

**Habitat**: Open sandy desert with some vegetation.

**Distribution**: Southwestern Iran, Arabia and North Africa and Sinai. In Israel, the subspecies *J. j. macrotarsus* is found in the northwestern Negev Desert.

**Taxonomic Remarks**: *J. macrotarsus* was first described as a species by Wagner (1843), 1840 according to Harrison and Bates (1991), from the Sinai Desert in the neighborhood northeast of the Suez Canal. Allen (1915) recorded *J. macrotarsus* from Wadi Fieran. Wassif (1953) noted it from El Arish; Maqdabah and Gebel Lehfan. Wassif and Hoogstraal (1954) recorded it from Wadi Raha, 5 km. west of St. Catherine's Monastery. Hoogstraal and Traub recorded it from Rumani and Wadi Feiran. The author found it 30 km. west of El Arish. It may intergrade with *J. j. schlueteri* in the northwestern Negev Desert or the north eastern Sinai Desert. Harrison (1992) believes *J. macrotarsus* is referable to *J. j. schlueteri*.

In view of the fact that *J. macrotarsus* cannot be distinguished from *J. jaculus*, but can be told from *J. j. schlueteri*, both morphologically and geographically, it is suggested that it be regarded as a subspecies of *J. jaculus* and be known as

*Jaculus jaculus macrotarsus* n. status

**Type locality**: Gebel-Libni, Sinai.

---

## Schlueter's Lesser Three-toed Jerboa  Plate 40

*Jaculus jaculus schlueteri* Nehring, 1901
Yarboa Matsui (Hebrew)
Jarboa, Djerba or Gaurti (Arabic)

**Description**: Head and body 98–125 mm.; ear 17–23 mm.; hind foot 58–67 mm.; tail 162–190 mm.

Schlueter's Lesser Three-toed Jerboa is a large subspecies of the Lesser Egyptian Jerboa, *J. j. jaculus*. It is distinctive for its long hind legs; pale pinkish ears; and black feather-like band near the end of the tail, which is divided lengthwise by a white streak on the underside.

The soles of the hind feet are hairy and buffy-white, and the vibrissae are mostly white. The upper parts are pale sandy-fawn, whitish about the eyes, on the sides and front of the muzzle, the flanks, and with a white band across the thigh. The underparts and tip of the tail are white.

**Habits**: Nocturnal and colonial, it digs long, shallow burrows. The gestation is 25 days, and the litter is 3–4. It feeds on succulent plants, such as saltwort and seeds, and is independent of free drinking water.

**Habitat**: Coastal sand dunes with some vegetation.

**Distribution**: Southwestern Iran, Arabian Peninsula and North Africa. The subspecies *J. j. schlueteri* is found along the southern coast of Israel, as far north as Jaffa. It may intergrade with *J. j. macrotarsus* n. status in the northwestern Negev and Sinai Deserts.

# DORMICE: GLIRIDAE

## DORMICE: MUSCARDINAE

Dormice are small squirrel-like rodents with long, furry tails. They are brownish or grayish in color. They live in trees, shrubs and among rocks. In cold weather, they become torpid. Confined to the Old World, they are represented in Israel by two species.

### Levant Garden Dormouse, Southwest Asian Garden Dormouse

Plate 41

*Eliomys quercinus melanurus* Ognev, 1940
Namneman Slaim (Hebrew)
Namnemon (Arabic)

**Description:** Head and body 107–130 mm.; ear 104–120 mm.; hind foot 23–26 mm.; tail 104–120 mm.

The Levant Garden Dormouse is believed to be a southern subspecies of the Eurasian Garden Dormouse *Eliomys quercinus*. It has pale grayish-brown upper parts and a pale buffy face with white cheeks. The eyes are spectacled with black, and the ears are large and pinkish. The underparts are white. It is distinctive for its tail, the terminal half of which is black and bushy.

**Habits:** It is nocturnal, inhabiting trees, shrubs and rocks. The nest is globular and compact, made of leaves and grass. It sometimes adapts a bird's nest for a foundation and adds to it. The nest is 0.8–3 meters above ground. After a gestation of 22–28 days, a litter of 2–7 pups is born. Fatty tissue is stored in the tail for winter hibernation. It is active as early as March and as late as September. Its diet consists of insects, small rodents, young birds, acorns, nuts, fruit and seeds.

**Habitat:** Forests, cultivated areas and rocky steppes.

**Distribution:** Europe and Asia. The subspecies *E. q. melanurus* is known from Asia Minor, Syria, Lebanon, Sinai and perhaps Africa. In Israel, it occurs on Mount Hermon, the Huleh Valley (Dan) and the Negev Desert (Wadi Naphekh).

**Taxonomic Remarks:** *Eliomys melanurus* Wagner, 1840, was described from Sinai. Its distribution is Southern Asia Minor and perhaps North Africa. It was considered by some authors to be a distinct monotypic species (Ellerman, 1941, 1951; Harrison, 1972). Other authors regarded it as a subspecies of *E. quercinus* L, 1766 (Ognev, 1940; Niethammer, 1959; Ranck, 1968). The author agrees with the latter. The measurements of *melanurus* fall within the range of *quercinus*. The features which distinguish it from its European relative;

the pale coloration; large ears; and large, tympanic bullae, are characteristic of races living on limestone in arid regions. The lack of white on the tip of the tail, typical of *E. quercinus* in Europe, is not absolute. In *E. quercinus* the underside of the tail varies from dirty white in Germany to blackish in Spain. A few white hairs have been found on the tip of the tail in a specimen of *E. melanurus* from Sinai (Harrison, 1972). A white tip on the tail can be quite variable intraspecifically, from a lot to none, as seen in the Bushy-tailed Jird, *Meriones (= Sekeetamys) calurus calurus*. As for the large, tympanic bullae and black terminal half of the tail, *E. quercinus superans* of eastern Europe also has large, tympanic bullae; pale gray-brown upper parts; and a dark ventral zone on the tail.

## Sooty Garden Dormouse                    Plate 41

*Eliomys quercinus fuscus* ssp. nov.

The extent of the variation in the color of *melanurus* is not known (Harrison, 1972). A series of eight specimens from the Golan Heights and six from Mount Hermon collected since 1967, which are in the collection of Tel Aviv University, and others at Beit Ussishkin, show that there are two substrate races living side by side: a pale form living on limestone on Mount Hermon; and a dark form inhabiting the basalt rocks and soil of the Golan Heights. Qumsiyeh (1996) reports two specimens from the H5 region at the Jordan University Natural History Museum as being distinct from each other in color, one being almost melanistic, and attributes this to marked polymorphism.

**Diagnosis**: The melanistic form represents a previously unknown substrate subspecies. It is the same size as *E. quercinus melanurus*. The upper parts are a dark sooty-gray, reflecting the dark volcanic soil it lives on. In addition to its general sooty coloration, the black of the spectacles, head markings and tail is more extensive, sometimes including the entire tail. The ears appear narrower, but otherwise it does not seem to differ in its cranial characters. A series of specimens from the Golan Heights and Jordan, reveal a population morphologically and geographically distinct from the type species. It is suggested that this new melanistic substrate race be known as

*Eliomys quercinus fuscus* ssp. nov.

**Holotype**: TAU M4689 female. Deposited in the Zoological Museum of Tel Aviv University.

**Type locality**: Bab El Hawa, Golan.

**Collector**: Aharon Keyser.

**Etymology**: Named after its sooty color.

## Forest Dormouse                    Plate 42

*Dryomys nitedula phrygius* Thomas, 1907
Namneman Etsim (Hebrew)
Namnemon (Arabic)

**Description**: Head and body 79+–110 mm.; ear 12–13 mm.; hind foot 19.5–20 mm.; tail 165+–105 mm.

A slightly smaller and paler race of the Forest Dormouse of Eurasia, it is distinguished by its light buffy, yellowish-brown upper parts; wide black mask that extends across the eyes and backwards to the base of the ears; and long, flattened, thickly-haired tail, uniformly colored blackish-brown. The cheeks, sides of the muzzle, feet and underparts are white.

**Habits**: Nocturnal and arboreal, it builds a leafy, rotund nest in oaks or low, thick bushes. It curls up like a ball when dormant for short periods in the winter. Gestation is 23–25 days. There are 1–4 young. It is omnivorous, eating fruits, buds, seeds, insects, arachnids and birds' eggs.

**Habitat**: Mountains with dense forests and thickets with low bushes and vegetation.

**Distribution**: Widespread across southeast Europe, Asia Minor, Arabia and as far east as India. The subspecies *D. n. phrygius* is known from Asia Minor, Syria and probably Lebanon. In Israel, it occurs only in Upper Galilee.

# MOLE RATS: SPALACIDAE

Mole Rats are small- to medium-sized mole-like rodents. Like a mole, they have a broad head; a stocky body; and no visible eyes or external ears. Unlike a mole, they have large, rat-like incisors; a vestigial tail; comparatively small claws on the forefeet; and burrow with their noses. They are gray or golden in color. They are strictly terrestrial and fossorial. Confined to the Old World, they are widespread. In Israel, they are represented by a single species.

## Greater Mole Rat, Blind Mole Rat    Plate 43

*Spalax microphthalmus ehrenbergi* Nehring, 1898 n. status

Holed (Hebrew)

Khlunt or Abu Amma (Arabic)

**Description**: Head and body 151–230 mm.; hind foot 17.8–22.3 mm.

The Greater Mole Rat of Israel is a small southern subspecies. It is easily recognized by its sausage-shaped body. The coloration is variable, ashy-gray, washed more or less with a golden sheen on the upper parts. The blunt head is broad and flat, with rudimentary eyes completely covered by skin; a blunt, pink nose; and a white streak from the nostril to the forehead and white stripes from the sides of the nose backwards towards the ears. The nose is protected by a horny pad. Its huge, rat-like incisors project even when the mouth is closed. There are no external ears or tail. Its hearing is highly sensitive. The Greater Mole Rat is the most specialized mammal in Israel.

**Habits**: Fossorial, it is highly adapted to subterranean life and burrows mostly with its head. It hardly ever appears above ground except during the mating season, spending much of its solitary life underground. Its burrows are exceptionally long and complicated. It communicates by tapping its head against the roof of the burrow. Its presence is often detected by a row of small mounds of excavated earth spaced somewhat evenly in a line about 1.2 meters apart. The gestation is about 30 days, and the litter is 3–4. It feeds on subterranean parts of plants, such as roots, tubers and bulbs, and stores them in a cache.

**Habitat**: Plains, cultivated fields, hilly steppes and mountain clearings, wherever the soil is suitable for digging. It is not found in true desert.

**Distribution**: Libya, Egypt and Sinai, eastern Europe, Asia Minor, southern Russia and Arabia. The subspecies *S. m. ehrenbergi* is known from Iraq, Syria, Lebanon, Jordan. In Israel, it is widespread on Mount Hermon, the Golan Heights, from the Galilee to the northern Negev Desert. Four sibling species (chromosomal forms) interbreed and hybridize in Israel. (Wahrman, Goiten and Nevo, 1969).

**Taxonomic Remarks**: *Spalax (Nanospalax) leucodon* cannot be morphologically and reliably distinguished from *S. microphthalmus*, except for two small foramina near the back of the skull in dorsal view (Macdonald and Barrett, 1993). Harrison (1972) makes no mention of these small openings in his anatomical description of the cranium of *S. leucodon*, (although they are present) and notes that it is extremely doubtful if *leucodon* is specifically distinct from *microphthalmus*. The two forms replace each other geographically in Europe where they may hybridize in Rumania. *Spalax leucodon* is best considered a monospecific, smaller subspecies of *S. microphthalmus* and part of a size and chromosomal cline running from north to south. It is suggested, therefore, that *S. leucodon ehrenbergi* be known as:

*Spalax microphthalmus ehrenbergi* n. status.

# MICE, OLD WORLD RATS AND HAMSTERS: MURIDAE

Field or wood mice are characterized by their prominent eyes; large ears; fairly long hind legs; and sparsely-haired, long tail. Hamsters are more compact, with small ears; short legs and tail; and have large cheek pouches. Old world rats and mice include the well-known species that are commensal with human beings, as well as the spiny mice, whose backs are covered with spine-like hairs, and the Bandicoot Rat. Field mice, hamsters and Old World mice and rats are native to the northern hemisphere of the Old World. In Israel, the Muridae are represented by three species of field mice, one species of hamster, two species of European rats, two species of house mice, two species of spiny mice and one species of Bandicoot Rat.

## OLD WORLD RATS AND MICE: MURINAE

### Broad-toothed Field Mouse, Big Levantine Field Mouse, Rock Mouse

Plate 45

*Apodemus mystacinus mystacinus* Danford and Alston, 1877

Ya'aron Gadol (Hebrew)

Far or Jeraidy (Arabic)

**Description**: Head and body 128–150 mm.; ear 16.8–20.5 mm.; hind foot 26–29 mm.; tail 115–146 mm.

The typical Broad-toothed Field Mouse of southeast Europe and Asia Minor is characterized by its large size; prominent eyes; large ears; fairly long hind legs; and sparsely-haired tail. The upper parts are grayish-brown, more or less darker brown or blackish on the back and top of the head. The underparts and feet are white. The tail is darker on the upper side. It may be distinguished from a young Common Rat by its comparatively short tail, less than the head and body combined.

**Habits**: It is mostly nocturnal. An active jumper rather than a climber, it occupies deep burrows or hollow trees. The nest is made of shredded grass and leaves. After a gestation of 21–29 days, usually 4–6 young are born. It feeds on plant material, grains, seeds, roots, berries and insects.

**Habitat**: From sea level to moister hills and mountains as high as 4,000 m. It inhabits rocky terraces, scrub land, forests and bush.

**Distribution**: Greece, Yugoslavia, Crete, Iraq, Lebanon and Jordan. In Israel, the subspecies *A. m. mystacinus* is found in the northernmost

brown ones in normal commensal conditions. The juvenile is grayish.

**Habits**: Mostly nocturnal, it climbs and swims well. The nest is made of shredded material. The gestation period is 18–20 days, and the litter is usually 4–7, but may reach 12. Prolific, it can reach plague proportions. Its diet consists of almost anything, including glue and soap. In the wild, it eats plants and seeds, insects and some meat. It causes damage to food and household articles and transports a host of diseases.

**Habitat**: Fields, gardens and desert, in and around human habitation.

**Distribution**: Cosmopolitan. The subspecies *M. musculus praetextus* is known from Syria, Lebanon, Jordan, Iraq, Saudi Arabia and Egypt (Sinai). In Israel, it is found mostly in cities and settlements throughout the country.

---

## Macedonian House Mouse                Plate 44

*Mus macedonicus* Petrov and Ruzic, 1983
Achbar HaBayit (Hebrew)
Far (Arabic)

**Description**: A sibling species, similar in size to the common House Mouse, *M. musculus praetextus*, except for its relatively shorter tail, relatively broad anterior part of the malar process, and biochemistry. In the desert, they are yellowish-brown.

**Habits**: Not known to differ from the common House Mouse.

**Habitat**: Feral, not commensal.

**Distribution**: Yugoslavia, Greece, Bulgaria, Turkey, Iran and Cyprus. In Israel, it is found in the Mediterranean Zone.

---

## Egyptian Spiny Mouse                Plate 46

*Acomys cahirinus dimidiatus* Cretzschmar, 1826
Kotsan Matsui (Hebrew)
Goup or Shoeshabgoop (Hadendowah Arabic)

**Description**: Head and body 80–118 mm.; ear 16.3–24 mm.; hind foot 17–20.5 mm.; tail 81–130 mm.

The typical Spiny Mouse is fairly small, although robust, and distinguished by fine, spine-like hairs that cover the hind part of the back. It has very large ears, and a long, scaly tail, subequal or longer than the head and body. It is highly variable in color. The upper parts are generally light buffy-brown, darker and grayer on the back, and contrasting sharply with the white underparts. There is a white spot below the eye and at the posterior base of the ear. The feet are white with pink, naked soles. It is darker in more humid areas, and paler in arid areas, strongly reflecting substrate coloration. The juvenile is grayer.

**Habits**: It is nocturnal and crepuscular. It inhabits niches and crevices among boulders and is very agile. In sandy areas, it uses gerbil burrows. The gestation is about 42 days, and the number of young 1–5. Although omnivorous, it eats mainly plant material, especially grain.

**Habitat**: Various rocky places, steppe-like desert, hills and sandy desert.

**Distribution**: Ranges from southern Iran, southern Asia Minor and Cyprus, Arabia, Jordan, Egypt, southern Algeria, south to Tanzania and west to Niger. In Israel, the subspecies *A. c. dimidiatus* is widespread from Galilee to the coastal plain, Mount Carmel, the southern Judean hills (around Jerusalem), near the Dead Sea and the Negev Desert.

## Southern Spiny Mouse                Plate 46

*Acomys cahirinus homericus* Thomas, 1923
Katsan Matsui (Hebrew)
Goup or Shoeshabgoop (Hadenowah Arabic)

**Description**: About the same size as the typical race.

The Southern Spiny Mouse is distinguished by its strikingly dark coloration. It is basically a buffy-brown, darker and grayer on the back and head, and lighter on the flanks and cheeks. The muzzle is pink, and the white spot is present at the base of the ear, but lacking below the eye. The mouth and throat are whitish, as are the toes, but the rest of the underparts are a dull pinkish-gray. The ears and upper side of the tail are dark gray.

**Habits**: Similar to typical subspecies.

**Habitat**: Inhabits basalt soil and rocks in moist thickets.

**Distribution**: Ranges from southern Iran, southern Asia Minor and Cyprus, Arabia, Egypt, Jordan, south to Tanzania and west to Niger and southern Algeria. The subspecies *A. c. homericus* occurs mainly in southern Arabia and Oman, but apparently is distributed in a kind of mosaic influenced by the dark substrate of soil and rocks. In Israel, it is found on the Golan Heights (near Kibbutz Sneer).

## Golden Spiny Mouse                Plate 46

*Acomys russatus russatus* Wagner, 1840
Kotsan Zahov (Hebrew)
Goup or Shoeshabgoop (Hadendowah Arabic)

**Description**: Head and body 100–115 mm.; ear 13–18 mm.; hind foot 15–19 mm.; tail 57–75 mm.

The typical subspecies of Golden Spiny Mouse is fairly small, though robust. It is distinctive for its short ears and small feet. It has coarse, spine-like hairs which are yellowish or grayish-rufous in color and cover the entire upper parts, almost up to the neck. The back is not noticeably darker than the sides. There is a white spot below the eyes and at the hind base of the ears. The nose, sides of the muzzle and ears are blackish. The underparts are buffy-white. There is no sharp demarcation between the flanks and the underparts. The feet are buffy-white with black soles. The bare tail is shorter than the head and body and black throughout.

**Habits**: It is active mainly in the early morning

and late afternoon. It inhabits niches and crevices in rocks. The gestation period is about 42 days, and the number of young 1–5, averaging three. Omnivorous, it feeds mainly on plant material.

**Habitat**: Stony desert hillsides and wadis with vegetation.

**Distribution**: Egypt, Syria, Saudi Arabia and South Yemen. In Israel, the subspecies *A. r. russatus* is found from the western side of the Dead Sea (Ain Faschka), south through the Negev Desert to Eilat.

**Taxonomic Remarks**: The subspecies *A. r. harrisoni* Atallah, 1970, was described from the western shore of the Dead Sea. Harrison (1972) notes that "the status of the population on the west shore of the Dead Sea in Israel is uncertain, possibly representing *A. r. harrisoni.*" He also states that "the material available is scarcely adequate to assess the full degree of individual variation in the species." The distinctive characters of *A. r. harrisoni*, of smaller size and paler color, is based on only two specimens. Qumsiyeh (1996) states that the continuous range and the few differences between *A. r. harrisoni* and *A. r. russatus* do not justify subspecies recognition.

---

**Short-tailed Bandicoot Rat**          Plate 44

*Nesokia indica bacheri* Nehring, 1897
Nesokia (Hebrew)
Girdi (Arabic)

**Description**: Head and body 202 mm.; ear 20 mm.; hind foot 38 mm.; tail 131 mm.

The Short-tailed Bandicoot Rat is a local race of the Asiatic Bandicoot Rat. It is the only species of its genus. It may be distinguished by its large size and stout, compact build; a thick head and a deep muzzle; short ears; a distinct, white gular spot; and short, sparsely-haired tail. The upper parts are pale yellowish or fawn-brown, becoming darker on the back, and mixed with longer, blackish hairs. The forefeet are whitish. The underparts are dusky yellowish-white, without a distinctive line of demarcation. The juvenile is sooty-gray.

**Habits**: Mostly nocturnal and fossorial, it burrows from 150–600 mm. below the ground. Sometimes it makes well-defined runways on the surface. There are 2–8 young born after a gestation of 21 days. Its diet consists of grass, roots and grain, which is sometimes stored in burrows. These are plugged with loose dirt.

**Habitat**: Desert oases, salt marshes, among sedge and shrubs near water, cultivated areas and banks of irrigation channels.

**Distribution**: Egypt, Syria, northern Arabia, Iran, Iraq, Afghanistan, western Pakistan, northern India, Russian and Chinese Turkestan. The subspecies *N. i. bacheri* is known only from Jordan (southeast of the Dead Sea) and Israel (Sdom and Ein Bedda), the Negev Desert (Ein Avdat), the Arava Valley (Ein Yahav) to Eilat.

# HAMSTERS: CRICETINAE

Hamsters are small- to medium-sized rodents, characterized by their large cheek pouches and short tails. They range from southeastern Europe, through Asia Minor, to northern Asia (southern Siberia). They are represented in Israel by a single species.

## Gray Hamster                                           Plate 43

*Cricetulus migratorious cinerascens* Wagner, 1848

Ogair (Hebrew)

**Description**: Head and body 82–113 mm.; ear 15–20.2 mm.; hind foot 14.1–17 mm.; tail 19–37 mm.

The Middle Eastern race of the Gray Hamster is distinguished by its small size and stocky appearance. It has large eyes; short ears; short legs with small toes; and a very short tail. The grayish-brown upper parts are slightly darker on the back and distinctly separated from the white underparts on the flanks and above the legs. The ears are blackish. The blunt muzzle and limbs are white. The thinly-haired tail is whitish-pink. The juvenile is grayer.

**Habits**: Active both day and night in spring and summer, it hibernates in the winter intermittently in burrows which it digs. These contain chambers for storing food, nesting and other uses. After a gestation of 16–19 days, a litter of 4–6 young is born. The diet contains young shoots, seeds and beans. The cheek pouches are so big that when they are full, the head may comprise one-third of the body.

**Habitat**: Shallow holes and galleries on high, rock-strewn mountain slopes with short shrubs and vegetation; also near cultivated areas.

**Distribution**: Greece, eastwards through Asia Minor, Arabia, southern Russia, the Ukraine, the Caucasus, Transcaucasia, Russian Turkestan, Iran, Afghanistan, Baluchistan, Kashmir, Chinese Turkestan and southwest Siberia. The subspecies *C. m. cinerascens* is known from Syria, Iraq and Lebanon. In Israel, it reaches the southern limit of its range in the northern half of the country, on Mount Hermon, the Golan Heights, Upper Galilee and the Mediterranean region.

# GERBILS, JIRDS AND SAND RATS: GERBILLINAE

Gerbils are mouse-sized rodents. They resemble jerboas in appearance and habits somewhat, although they are smaller, with shorter hind legs, a relatively longer muzzle and five functional toes on each foot. Jirds and sand rats are somewhere between gerbils and rats in size, with large eyes, hairy tails and whitish underparts. They are all nocturnal and fossorial, inhabiting arid steppes, deserts and coastal sand dunes. The gestation period is 20–21 days, and the litter ranges from 1–7. Their diet consists of roots, nuts, grass and insects. Confined to the Old World, the Gerbillinae probably surpass all other rodents in species and numbers. In Israel, they are represented by six species of gerbils, five species of jirds and one species of sand rat.

## Pygmy Gerbil                              Plate 47

*Gerbillus henleyi mariae,*Bonhote, 1909

Gerbeel Zair (Hebrew)

Baiyondeh (Arabic)

**Description**: Head and body 53–80 mm.; ear 8–10 mm.; hind foot 18–20 mm.; tail 73–92 mm.

The Pygmy Gerbil is distinguished by its very small size. The upper parts are a very pale sandy with a light grayish tint, especially on the back. There is a white spot above the eyes and at the base of the ear. The underparts are white, sharply demarcated on the flanks. The small feet have naked soles. The tail is darker towards the end but without a tuft. Despite its small size, it can be distinguished from the juveniles of other species of gerbils by its adult proportions.

**Habits**: It is nocturnal. It burrows beneath bushes in sand dunes and feeds mainly on seeds. Gestation is 19 days, and the litter is 3–6.

**Habitat**: Desert wadis with sandy hummocks covered by coarse vegetation; sand dunes, pebbly desert and coastal desert with nearly complete plant cover.

**Distribution**: North African Sahara, from Algeria through Libya, Egypt and northwestern Arabia. The subspecies *G. h. mariae* is known from Sinai, Israel and Jordan. In Israel, it has been found in the northern and central Negev Desert, practically to Eilat.

## Baluchistan Gerbil                        Plate 47

*Gerbillus nanus arabium* Thomas, 1918

Gerbeel Arava (Hebrew)

Baiyondeh (Arabic)

**Description**: Head and body 67–90 mm.; ear 8–14.5 mm.; hind foot 20–25.3 mm.; tail 80–145 mm.

The Baluchistan Gerbil of northwestern Arabia is distinguished by its soft, dull fawn upper parts, becoming darker on the back, and with white hairs at the base of the rump. There is a trace of a dark streak on the top of the nose, and a white spot over the eyes and behind the base of the ears. The underparts, legs and feet are white, and the soles of the feet are naked but have sparse hairs. At the end of the tail, there is a slight, pale-colored tuft.

**Habits:** Nocturnal and fossorial, it burrows in colonies beneath bushes and shrubs. The litter is 2–5.

**Habitat:** Sandy deserts, on dunes adjacent to as well as on salt flats and beds of wadis.

**Distribution:** Baluchistan, Arabia and Egypt. The subspecies *G. n. arabium* is known from northwestern Arabia, southwestern Iraq, Oman, South Yemen, Saudi Arabia, Kuwait, Jordan and Egypt (Sinai). In Israel, it is widespread in the Negev Desert, from the southern end of the Dead Sea (Sdom), south through the Arava Valley to Eilat.

## Wagner's Gerbil Plate 47

*Gerbillus dasyurus dasyurus* Wagner, 1842
Gerbeel Sla'im (Hebrew)
Baiyondeh (Arabic)

**Description:** Head and body 68–99 mm.; ear 11–15.5 mm.; hind foot 20–26.5 mm.; tail 81–128 mm.

The typical Wagner's Gerbil of northern Arabia closely resembles the Baluchistan Gerbil in its size and general pale color, sandy-buff above and white below. It may be distinguished by its lack of a dark nasal streak; indistinct, whitish spot over the eyes; dusky-black ears;

completely naked soles on its feet; and a blackish tuft at the end of the tail.

**Habits:** It is nocturnal and not gregarious. Fossorial, it burrows and frequents holes under loose rock slabs. It feeds on grasses and seeds.

**Habitat:** Rocky steppe desert, wadis, silt dunes and hammada. Gestation is 24–26 days, and the litter is 2–6.

**Distribution:** Arabia, Egypt (Sinai) and possibly Africa. The substrate subspecies *G. d. dasyurus* is known from Syria, Iraq, Saudi Arabia and Jordan. In Israel, it is found in the southern half of the country from the northwest end of the Dead Sea and south from Beersheba.

## Lehman's Gerbil Plate 47

*Gerbillus dasyurus leosollicitus* Von Lehmann, 1966
Gerbeel Sla'im (Hebrew)
Baiyondeh (Arabic)

**Description:** Head and body 83–191 mm.; ear 11.8–13.2 mm.; hind foot 21.8–24 mm.; tail 108–118 mm.

Lehman's Gerbil is a substrate subspecies of *G. dasyurus*, distinctive for its darker coloration. The upper parts are a dark brown with black speckling, more intense on the back and rump. The tuft at the end of the tail is black.

**Habits:** Similar to the nominate race.

**Habitat:** Mediterranean littoral hills.

Distribution: Arabia, Egypt (Sinai) and possibly Africa. The subspecies *G. d. leosollicitus* is known from Syria and probably Lebanon. In Israel, it is found in the northern half of the country, from Upper Galilee (Rosh Hanikra), Mount Carmel (Haifa), the coastal plain (Wadi Ara) and the Judean Hills (Jerusalem).

## Anderson's Gerbil                    Plate 47

*Gerbillus andersoni bonhotei* Thomas, 1919
Gerbeel Hof (Hebrew)
Baiyondeh (Arabic)

Description: Head and body 93–97 mm.; ear 14.5–16 mm.; hind foot 26–30 mm.; tail 112–125 mm.

Anderson's Gerbil is distinguished by its bright, sandy-colored upper parts, slightly darker on the back. It has a large, white spot above the eyes and behind the base of the ears. The ears are rather large, and the upper inner part is blackish. The underparts are white, sharply demarcated on the sides. The legs and feet are white, and the soles are mostly hairy, with a small, naked patch on the heel. The tail is sandy on the upper side and white below, with a thin, blackish dorsal line towards the tip, which is not tufted.

Habits: It is a nocturnal psammophile, burrowing beneath bushes and shrubs. It feeds on grasses and seeds.

Habitat: Desert wadis and sand dunes with some vegetation.

Distribution: Libya, Egypt and Jordan. The subspecies *G. a. bonhotei* is found in the northern costal plain of Sinai and Jordan. In Israel, it occurs in the southern coastal plain and northwestern Negev Desert. It intergrades with *G. a. allenbyi* between Ashkelon and Kerem Shalom.

## Allenby's Gerbil                    Plate 47

*Gerbillus andersoni allenbyi* Thomas, 1918
Gerbeel Hof (Hebrew)
Baiyondeh (Arabic)

Description: Head and body 70–111 mm.; ear 9–16 mm.; hind foot 24–29 mm.; tail 95–130 mm.

Allenby's Gerbil is a northern race of Anderson's Gerbil. It is distinguished by its darker sandy-buff upper parts, which are darker on the back. It has a white spot above the eyes and at the hind base of the ears. The ears are rather large and dusky on the inner sides. The juvenile is a dull grayish-brown. The underparts are white, and the soles are mostly hairy with a small, naked patch on the heel. The upper side of the tail is darker, and becomes grayish towards the tip, which is virtually not tufted.

Habits: It is a nocturnal psammophile, burrowing beneath shrubs and bushes. It feeds on grasses and seeds. Gestation is about 21 days and the litter is 3–5.

Habitat: Coastal sand dunes, sandy soils and sandstone hills with some vegetation.

Distribution: Libya, Egypt and Jordan. The

subspecies *G. a. allenbyi* is endemic in Israel, where it is confined to the narrow littoral zone of the Mediterranean, from Haifa south to Ashkelon, where it intergrades with the more richly colored *G. a. bonhotei*.

## Lesser Egyptian Gerbil                    Plate 47

*Gerbillus gerbillus asyutensis* Setzer, 1960
Gerbeel Dromi (Hebrew)
Baiyondeh (Arabic)

**Description**: Head and body 75–87 mm.; ear 11–13 mm.; hind foot 27–30 mm.; tail 115–120 mm.

The Lesser Egyptian Gerbil of North Africa is small, but has long hind legs and a long tail. The upper parts are a rich reddish-fawn, with fine, black speckling which is absent in the typical race. The area around the eyes is whitish, and the ears are pinkish-white, with a white spot at the hind base. The underparts are white, sharply demarcated along the sides. The legs and feet are white, and the soles hairy. The tail is weakly bicolored, pale buff above and white underneath. The terminal tuft is grayish, lacking the narrow, black dorsum line in the typical form.

**Habits**: It is nocturnal. A colonial psammophile, it burrows in sand to a depth of a meter or more. The entrances are usually blocked with loose sand. When alarmed, it can jump by leaps and bounds, up to 1.5 meters high. After a gestation period of 20–21 days, a litter of 2–4 or more is born. Its diet consists of grass, roots, nuts and insects.

**Habitat**: Sandy desert and sandy littoral regions with sparse vegetation.

**Distribution**: Africa, from Egypt eastwards to Iraq and Iran, south through the Arabian Peninsula and Sinai (Nabeq). The subspecies *G. g. asutensis* is known from Upper Egypt (southeast of Asyut, eastern desert). In Israel, it has been recorded from the northwestern Negev Desert and is slightly larger in the Arava Valley.

**Taxonomic Remarks**: Cheesman's Gerbil, *G. cheesmani* Thomas, 1919, Plate 47, is distinguished from *G. gerbillus* mainly on the basis of its large tympanic bullae which projects beyond the supraoccipital. This character in desert forms is usually associated with subspecific rather than specific status. The teeth of *G. cheesmani* are basically the same in size and structure as those of *G. gerbillus*. The size of the head and body, ear, hind foot and tail of *G. cheesmani* overlaps those of *G. gerbillus*. The two forms are not sympatric and replace each other geographically. It is suggested, therefore, that *G. cheesmani* Thomas, 1919, cannot be sustained as a distinct species, and that it be subsumed under *G. gerbillus* Olivier, 1801, which has name priority, and be known as

*Gerbillus gerbillus cheesmani* n. status.

Harrison (1972) states that *G. cheesmani* of eastern Arabia is possibly separated from *G. gerbillus* by the mountains of the Jordan and the Hejaz. In Israel, the population of *G. gerbillus* in the northwestern Negev Desert is disjunctive with the population in the Arava Valley, which is somewhat larger, and may belong to the subspecies *G. g. cheesmani* or an intermediate population.

## Greater Egyptian Gerbil                    Plate 46

*Gerbillus pyramidum floweri* Thomas, 1919

Gerbeel Holot (Hebrew)

Dimsi (Arabic)

**Description:** Head and body 95–130 mm.; ear 14–16 mm; hind foot 30.5–38 mm.; tail 130–149 mm.

Although somewhat smaller than the typical Greater Egyptian Gerbil of North Africa, the subspecies *G. p. floweri* resembles the Lesser Egyptian Gerbil, except that it is larger. The upper parts are a paler or duller sandy-buff, washed with blackish-tipped hairs on top of the head and back. There is a whitish spot above the eyes, and it has whitish cheeks. The ears are pinkish-white. The underparts are white, sharply demarcated along the flanks and side of the neck. The legs and feet are whitish and less hairy than in the Lesser Egyptian Gerbil. The tail is bicolored, sandy-buff above and white beneath, and the terminal tuft is a mixture of grayish-white and brown.

**Habits:** It is nocturnal. A psammophile, it burrows in colonies. The entrance holes are plugged with sand. The gestation period is approximately 22 days. The litter is usually four, but may be more. Its diet is mainly seeds and sometimes camel droppings. It can leap directly up to a height of 660 mm.

**Habitat:** Coastal dunes and sandy desert with sparse vegetation.

**Distribution:** North Africa, from Morocco eastwards to Egypt, and southwards to Asben and Sudan and northwestern Arabia. The subspecies *G. p. floweri* is known from the northern Sinai Desert (south of El Arish). In Israel, there is a morphologically indistinguishable chromosomal cline from the northern Negev Desert up the coastal plain to Holon (Wahrman and Zahavi, 1955). An allopatric population is found as far north as Acre.

## Tristram's Jird                    Plate 48

*Meriones tristrami tristrami* Thomas, 1892

Merion Matsui (Hebrew)

Gherda (Berber Arabic)

**Description:** Head and body 107–124 mm.; ear 12–19 mm.; hind foot 29.1–32 mm.; tail 127–144 mm.; tympanic bullae 11.2–13.3.

The typical Tristram's Jird of Asia Minor may be distinguished by its fairly small size and generally dull or bright fulvous upper parts, with or without variegated black on the back. It has relatively narrow, naked, dusky-gray ears, paler around and in front of the eyes and behind the ears. The ears are rather long and pinkish-gray. The underparts are white, sharply demarcated on the flanks, side of the neck and hind legs. The feet are white, and the soles are mostly hairy, with a naked streak extending to the heels. The tail is yellowish-fulvous, lightly grizzled with black above, paler without black below, and with a more or less blackish, slightly-tufted tip.

**Habits**: It is nocturnal. Fossorial, it burrows in the sides of hillocks and mounds. The nesting chamber is made of finely-divided plant material, sometimes with shredded paper or other refuse. The gestation is 25–29 days with a litter from 1–7. It feeds on grain, seeds, leaves and stems and may store food. This jird shows an unusual resistance to extreme temperatures and thirst.

**Habitat**: Deserts, steppes and cultivated fields. It occurs on a wide range of soils, including sand, alluvial deposits, loess and terra rosa.

**Distribution**: Asia Minor, Iran, Iraq, Syria, Turkey, Lebanon and Sinai (El Arish). In Israel, the subspecies *M. t. tristrami* is common in the north in the valleys, less common on the mountains, the coastal plain.

## Tristram's Syrian Jird           Plate 48

*Meriones tristrami bodenheimeri* Aharoni, 1932
Merion Matsui (Hebrew)
Gherda (Berber Arabic)

**Description**: Head and body 111–152 mm.; ear 16–22.5 mm.; hind foot 31–37 mm.; tail 111–160 mm.

The substrate subspecies *Meriones tristrami bodenheimeri* is distinguished by its larger size, on average, and darker coloration. The upper parts are a dark gray, sprinkled with yellow-

brown. The tail tip is black and almost never white.

**Habits**: Similar to the typical race.

**Habitat**: Basalt rocky steppes with black soil.

**Distribution**: Asia Minor, Iran, Iraq, Syria, Turkey, Lebanon and Sinai (El Arish). The subspecies *M. t. bodenheimeri* is known from Syria and Lebanon. In Israel, it is found on the Golan Heights.

## Tristram's Desert Jird           Plate 48

*Meriones tristrami deserti* ssp. nov.
Merion Matsui (Hebrew)
Gherda (Berber Arabic)

**Description**: Head and body 107–138 mm.; ear 17–18 mm.; hind foot 29–30 mm.; tail 98–138 mm.; length of skull 39+ mm.; tympanic bullae 12.5–14 mm.

The upper parts are pale sandy-fawn with little red or black; the underparts are white. The tail has a blackish tip.

**Habits**: Probably the same as previous races.

**Habitat**: Semi-desert

**Distribution**: A previously unrecognized subspecies of Tristram's Jird inhabits the northern Negev Desert and northern coast of the Sinai Desert.

**Diagnosis**: It is distinguished from the nominate race to the north by its paler coloring and inflated, laterally-protruding tympanic bullae give a squarish shape to the occipital region in dorsal aspect.

In view of this population of Tristram's Jird's adaptation to the arid conditions of the desert, and its geographical distribution, it is suggested that it be known as

*Meriones tristrami deserti* ssp. nov.

**Holotype**: TAU M2844 male

**Paratypes**: M4542 El Arish: Head and body 115 mm.; tail 120 mm.; hind foot 29 mm.; ear 17.5 mm.

M7131 5 km. south of Beersheba: Head and body 138 mm.; tail 98 mm.; hind foot 29 mm.; ear 18 mm.

**Type locality**: 5 km. south of Beersheba

**Etymology**: Named after the desert it inhabits.

Deposited in the Zoological Museum of Tel Aviv University.

**Sundevall's Jird**      **Plate 48**

*Meriones crassus crassus* Sundevall, 1842
Merion Midbar (Hebrew)
Gherda (Berber Arabic)

**Description**: Head and body 112–140 mm.; ear 14.5–18 mm.; hind foot 27–35 mm.; tail 110–150 mm.

The typical Sundevall's Jird of North Africa is distinguished by its smallish size and sandy-buff upper parts, slightly paler around the eyes and at the hind base of the ears, which are small and pale gray. The underparts are white with a fairly distinct demarcation along the flanks.

The legs and feet are white. The soles of the hind feet are partially hairy. The tail is relatively short and thick, about the same length as the head and body or slightly shorter, sandy-gray with a black terminal crest about one third the length of the tail.

**Habits**: It is nocturnal. Colonial and fossorial, it digs elaborate burrow systems up to a depth of 1.5 meters. The gestation period is 22–24 days, and the litter is from 4–7. Its diet consists mainly of seeds but also locusts. It appears independent of free drinking water.

**Habitat**: Sandy deserts with hillocks and little vegetation, and limestone steppes, overblown with sand.

**Distribution**: North Africa from Morocco, east to Egypt and south to Asben and Sudan, throughout Arabia, Iran, southern Russian Turkestan, Afghanistan and Waziristan. The typical subspecies *M. c. crassus* is known from Egypt (Sinai), Jordan, northern and central Arabia, Saudi Arabia, Kuwait and Oman. In Israel, it is found in the southern half of the country, in the Negev Desert and Arava Valley, south to Eilat.

**Buxton's Jird, Negev Jird**      **Plate 48**

*Meriones lybicus sacramenti* Thomas, 1922
n. status
Merion Holot (Hebrew)
Gherda (Berber Arabic)

**Description**: Head and body 135–190 mm.; ear

13–22 mm.; hind foot 33–41 mm.; tail 110–168 mm.

Buxton's Jird is distinguished by its large size, small ears and relatively short tail. The upper parts are sandy-buff, diffusely speckled with black, becoming paler and clearer on the flanks, above and before the eyes, and whitish behind the base of the ears. The ears are dusky-pink. The underparts are white, fairly well demarcated on the flanks. The legs and feet are white. The soles of the hind feet are partly hairy. The tail is grayish-buff, with a distinctive black terminal tuft and crest that extends as a black line to its base, and may have a white tip. Individuals from the coastal dunes are darker above, whereas those from the Negev Desert are lighter.

**Habits:** It is nocturnal. A psammophile, it burrows in sand dunes. Gestation is about 24–25 days, and the litter is 2–8.

**Habitat:** Sandy desert, shifting sand dunes and more or less stable sand formations.

**Distribution:** Confined to Israel, there are two populations, a slightly larger one in the coastal plain as far north as the Yarkon River; the other from the Negev Desert (area of Beersheba).

**Taxonomic Remarks:** *M. sacramenti* is claimed to be the only species of mammal endemic to Israel (Mendelssohn and Yom-Tov, 1999). Its distribution is allopatric and microgeographic for a species of jird. It is also curious that *M. lybicus* Lichtenstein, 1832, is found on both sides of Israel, but not in Israel. *M. sacramenti*

replaces *M. lybicus* in Israel geographically and is nowhere sympatric.

*M. sacramenti* and *M. lybicus* are considered distinct species, although they resemble each other in most of their anatomical and dental characters and are close cytologically. *M. sacramenti* has 46 chromosomes. *M. lybicus* has 44 chromosomes. They differ only in characters that intergrade marginally and overlap without a morphological gap between them. The size of the head and body and tympanic bullae of *M. sacramenti* are larger on average than in *M. lybicus*, but this is not diagnostic of distinct species, rather of subspecies. It is therefore suggested that *M. sacramenti* be considered as a local subspecies of *M. lybicus*, and be known as:

*Meriones lybicus sacramenti* n. status.

**Bushy-tailed Jird**      **Plate 48**

*Meriones (Sekeetamys) calurus calurus* Thomas, 1892
Yefet Zanav (Hebrew)
Gherda (Arabic)

**Description:** Head and body 122–126 mm.; ear 17–19 mm.; hind foot 30.6–33 mm.; tail 112–148 mm.

The Bushy-tailed Jird is unique among the Gerbillinae in having a bushy tail. It is of medium size, with reddish-buff upper parts, faintly speckled with black and darker on the back. There is an indistinct, whitish spot above the eye and one behind the base of the ears. The ears are rather large and a dusky pinkish-gray. The underparts are white, fairly well-demarcated along the flanks. The legs and feet are white, and the soles of the hind feet are naked. The dark, bushy tail is usually longer than the head and body and may have a white tip, which varies from half the tail to none at all.

**Habits**: It is nocturnal. Fossorial, it burrows under boulders and rocky ledges. It is also a nimble climber. Gestation is 24 days and the litter is 2–7. Its diet contains dry vegetation and seeds. Otherwise, little is known.

**Habitat**: Barren rocky deserts, on mountainsides with sparse vegetation.

**Distribution**: Eastern Egypt, Sinai and Jordan. The typical subspecies is found in Sinai and Jordan. In Israel, it inhabits the western side of the Dead Sea, south through the eastern and southern Negev Desert, as far as Eilat.

**Taxonomic Remarks**: The Bushy-tailed Jird was earlier regarded as *Meriones calurus* (Walker, 1964), but then considered unique and placed in a monotypic genus or subgenus *Sekeetamys* on the basis of its bushy tail and ecology. It is, in fact, not all that unique. In the United States, the genus *Neotoma* consists of several species of woodrats with hairy, not scaly tails, but one of them, *Neotoma cinerea*, has a bushy tail. It is not regarded as belonging in a separate genus. Furthermore, the juvenile Bushy-tailed Jird does not always have a bushy tail. It develops with age. Where then does it belong? Chromosome banding studies clearly show that this species is closely related to the genus *Meriones* (Qumsiyeh, 1996). It is suggested that the Bushy-tailed Jird be

regarded as a highly specialized species of the genus *Meriones*, and that the earlier name, *Meriones calurus calurus*, be reinstated.

---

## Fat Sand Rat                                    Plate 48

*Psammomys obesus terrae-sancta* Thomas, 1902

Psamon Midbar (Hebrew)

**Description**: Head and body 138–195 mm.; ear 13.5–17 mm.; hind foot 35–42 mm.; tail 95–152 mm.

The Fat Sand Rat is distinguished by its large size and robust appearance. It has small ears, thick feet and a short tail, less than the length of the head and body. The upper parts are generally a mixture of light rufous-brown, diffusely and inconspicuously speckled with black. The flanks are pale ochre, and it has yellowish or grayish-white underparts, legs and feet. The color varies, darker in more humid places, paler in arid ones. The soles of the feet are naked and dusky, with some longer hairs and a naked patch on the heels. The thick tail is the same color as the body, becoming black towards the slightly-tufted tip.

**Habits**: It is usually nocturnal, but often active during the day. It burrows colonially in warrens with large entrance holes, usually having excavated soil in front. It may be seen sunbathing or feeding, sometimes sitting upright like the Prairie Dog. Gestation is 24 days, and there are 1–7 young in a litter. Its diet consists of succulent plants and grains which it stores in chambers. It is independent of free water.

**Habitat**: Sandy desert, sometimes rocky terrain or low sandy mounds with vegetation.

**Distribution**: North Africa, from Algeria to Egypt, south to Sudan and east across Arabia. The subspecies *P. o. terrae-sancta* occurs in Sinai, Syria and Jordan. In Israel, it is found north and west of the Dead Sea, the central Negev Desert and Arava Valley, from south of Beersheba to Yotvata.

## VOLES: MICROTINAE

Voles are small, stocky rodents, characterized by their small, beady eyes; small ears; short legs; and short, hairy tails. They are generally brownish in color. They are fossorial and active during the day, as well as at night. Widely distributed in the northern hemisphere, they are represented in Israel by three species, one of which seems to have been extirpated.

### Syrian Water Vole                     Plate 43

*Arvicola terrestris hintoni* Aharoni, 1932

Navran Mayim (Hebrew)

Far (Arabic)

**Description**: Head and body 185–195 mm.; ear 12–17 mm.; hind foot 37–38 mm.; tail 125–130 mm.

The Water Voles of Eurasia are the largest voles of the Old World. The Syrian Water Vole may be distinguished by its large size and relatively long tail for a vole. Adapted for aquatic life, the fur is thick and wooly, and there is a hairy swimming fringe on the feet. The upper parts are dark brown, suffused with rusty-brown anteriorly. The underparts are dark grayish-white, fairly well demarcated on the flanks and cheeks. A long, brownish streak runs down the middle of the chest. The feet are grayish-white. The tail is well haired, light buffy-brown above and paler beneath.

**Habits**: It is active day or night. Colonial, it builds burrows that terminate beneath the water. Amphibious, it swims and dives well. It keeps well-defined runways that lead to the water. The nest is made of grass and other plant material, built slightly above water level, under logs, driftwood or dense vegetation. After a gestation period of about 42 days, 2–7 young are born. It feeds mainly on reeds and other aquatic plants, which are usually stored for the winter.

**Habitat**: Grassy banks of streams where the water level is constant. Sometimes in cultivated fields and wet meadows.

**Distribution**: Eurasia, Asia Minor and northern Arabia. *A. t. hintoni* is known from

Asia Minor, Turkey (Lake Antioch). In Israel, its presence is a mystery. It has been reported as common near the Banias, but the only specimens known are skulls found in owl pellets in the vicinity of Lake Huleh at Yessod Hama'ale and near Melaha.

---

## Hermon Snow Vole                    Plate 43

*Microtus nivalis hermonis* Miller, 1908
Navran Sheleg (Hebrew)
Far (Arabic)

**Description**: Head and body 96–137 mm.; ear. 13.6–18 mm.; hind foot 18.1–22 mm.; tail 52–70 mm.

The Hermon Snow Vole is distinguished by its small size. It has small eyes and ears, and pale coloration. Its tail is long, subequal with half the length of the head and body. The upper parts are light clay-brown, diffused with inconspicuous, black speckling on the back and shoulders, lighter on the cheeks and flanks. The ears are blackish. The underparts are grayish-white, indistinctly demarcated on the flanks. The feet are buffy-white, and the soles of the feet blackish. The tail is whitish. Some individuals are darker gray, and others bright reddish-gray on the lower back. The juvenile is gray.

**Habits**: Although claimed to be strictly nocturnal, it may be active during the day in the winter, when temperatures are not as cold, and the black of its ears that protect it against the sun's ultraviolet rays, is useless at night. It burrows beneath boulders. The gestation period is 21 days, and the short alpine summer allows only one or two litters of 4–7 young. It feeds on a variety of grasses and herbs and occasionally carrion.

**Habitat**: High mountains on rocky slopes and in gorges among boulders and scree above the tree line. In Iran, it ranges as high as 3,367 m.

**Distribution**: Europe, southwestern Turkestan, Iran, Asia Minor and northwestern Arabia. The subspecies *M. n. hermonis* is known from Lebanon. In Israel, it is found on Mount Hermon from 1,650 to just under 2,000 m. above sea level.

---

## Gunther's Social Vole, Levant Vole   Plate 43

*Microtus socialis guentheri* Danford and Alston, 1880
Navrat Sadot (Hebrew)
Far (Arabic)

**Description**: Head and body 92–140 mm.; ear 10–15 mm.; hind foot 15–22 mm.; tail 16–35 mm.

Gunther's Social Vole is distinguished by its small size. It has small eyes and ears, and a short tail, which is only one quarter of the length of the head and body. The upper parts are reddish or grayish-brown, grayer in the juvenile. The ears are short and covered with short, white hairs. The underparts are grayish-white, faintly buffy on the abdomen. The feet are pale buffy. The soles of the hind feet are densely hairy. The tail varies greatly in length; it is fawn brown above, paler beneath and slightly darker towards the tip.

**Habits**: Nocturnal and diurnal. Fossorial, and colonial, it burrows in cultivated areas, often near water. It is sometimes driven to the hills by heavy rains. The burrows are 5–8 cm. below the surface and are complicated with entrances, nesting and storage chambers. The gestation is 21 days, and the number of young 4–8, but on rare occasions as high as 18. Its diet consists of grasses, herbs, and can be a menace to winter crops. Fluctuation in numbers is extraordinary, sometimes resulting in a plague.

**Habitat**: Grassy terrain, cultivated fields, bushy scrub, on mountain slopes and valleys, in open oak forests and dry hillsides, from sea level to 2,000 m., and even above the tree line.

**Distribution**: Greece, Asia Minor to northern Arabia, Russia, Iran and Afghanistan, and Libya. The subspecies *M. s. guentheri* is found in Turkey, Syria and Lebanon. In Israel, it is widespread throughout the northern half of the country, south to Mishmar HaNegev.

The occurrence of Gunther's Social Vole in Israel was first discovered not in Israel, but in the British Museum when a specimen of the snake, *Caelepeltis lacertina*, collected by Tristram in 1863 on the Plain of Gennesaret, was found to contain a perfect specimen of Gunther's Social Vole in its stomach.

# REFERENCES

Allen, G.M., 1915. Mammals obtained by the Phillips Palestine Expedition. *Bull. Mus. Comp. Zool.* Cambridge, Mass., 59: 3–14.

Allen, G.M., 1939. Checklist of African Mammals, *Bull. Mus. Comp. Zool.* at Harvard, Vol. 83.

Anderson, J. and W.E. de Winton, 1902. *Zoology of Egypt: Mammalia*, Hugh Rees Pub. Ltd. London, pp. 374.

Atallah, S.L., 1977. The mammals of the eastern Mediterranean region: their ecology, systematics, and zoogeographical relationships (part 1), *Saugetierkd. Mitt.*, 25: 241–320.

Atallah, S.L., 1978. The mammals of the eastern Mediterranean region: their ecology, systematics and zoogeographical relationships (part 2). *Saugetierkd. Mitt.*, 26: 1–50.

Bodenheimer, F.S., 1935. *Animal Life in Palestine*. L. Mayer, Jerusalem xiii + 506 pp.

Bodenheimer, F.S., 1960. *Animal and Man in Bible Lands*. E.J. Brill Publ., Leiden, pp. 232.

Bodenheimer, F.S., 1958. The Present Taxonomic Status of the Terrestrial Mammals of Palestine. *Bull. Research Council of Israel* Section B: Zoology, October No. 2–4, Vol. 7B.

Catzeflis, F., T. Maddalena, S. Hellwing & P. Vogel, 1985. Unexpected findings on the taxonomic status of East Mediterranean *Crocidura russula* auct. (Mammalia, Insectivora). *Zeitschrift Saugetierkund.* 50, 185–201.

Dobson, G.E., 1878. *Catalogue of the Chiroptera in the Collection of the British Museum.* Reprinted 1966 and published by Verlag J. Cramer, German, xlii + 567 pp.

Dorst, J.& P. Dandelot, 1972. *A Field Guide to the Larger Mammals of Africa.* Collins, London. pp. 287.

Ellerman, J.R., & T.C.S. Morrison-Scott, 1951. *Checklist of Palearctic and Indian Mammals, 1758 to 1946.* British Museum (Nat. Hist.), London, vi + 810 pp.

Ferguson, W.W., *Guide to Mammals of our Land* [Hebrew]. Nature Reserves Authority, Tel Aviv. pp. 27.

Ferguson, W.W., *Living Animals of the Bible*, Charles Scribner's Sons, New York. pp. 95.

Ferguson, W.W., 1981 a. The systematic position of *Canis aureus lupaster* (Carnivora: Canidae) and the occurrence of *Canis lupus* in North Africa, Egypt and Sinai. *Mammalia*. 45: 459–465.

Ferguson, W.W., 1981 b. The systematic position of *Gazella dorcas* (Artiodactyla: Bovidae) in Israel and Sinai. *Mammalia*, 45: 453–457.

Ferguson, W.W., A. Shmida & M. Livneh, *On the Snow of Mount Hermon* [Hebrew]. Nature Reserves Authority and Ministry of Defense Publishing House, Tel Aviv. pp. 144.

Ferguson, W.W., Y. Porath & S. Paley. 1985. Late Bronze Period yields evidence of *Dama dama* (Artiodactyla: Cervidae) from Israel and Arabia. *Mammalia*, 49: 209–214.

Filippucci, M.G., S. Simson & E. Nevo. 1989. Evolutionary biology of the genus *Apodemus* Kaup,

1829 in Israel: Allozymic and biometric analyses with description of a new species: *Apodemus hermonensis* new species (Rodentia, Muridae). *Bull. Zool.*, 56: 361–376.

Flower, S.S., 1932. Notes on recent mammals of Egypt, with a list of species recorded from that kingdom. *Proc. Zool. Soc. Lond.*, 101: 369–450.

Gasperretti, J., D.L. Harrison and W. Buettiker, 1985. The Carnivora of Arabia in *Fauna of Saudi Arabia*. p. 405.

Golani, I. & A. Keller, 1975. A longitudinal field study of the behavior of a pair of golden jackals. pp. 303–335, in *The Wild Canids: Their Systematics, Behavioral Ecology and Evolution* (M.W. Fox, ed.) Van-Nostrand Reinhold, New York and London, pp. 508.

Goodwin, G.G., 1940. Mammals collected by the Legendre 1938 Iran Expedition. *Amer. Mus. Novit.*, No. 1082, pp.1–17.

Groves, C.P., 1983. Notes on gazelles IV. The Arabian gazelles collected by Hemprich and Ehrenberg. *Zeitschrift Saugetierkund.*, 48: 371–381.

Harper, F., 1945. *Extinct and Vanishing Mammals of the Old World*. Special Publication No. 12, Amer. Committee for the International Wild Life Protection, New York Zoological Park, N.Y. pp. 849.

Harrison, D.L., 1955. *The Mammals of Arabia*, Vol. I. Introduction, Chiroptera, Insectivora, Primates. Ernest Benn Ltd., Lond, xx + 192 pp.

Harrison, D.L., 1968. *The Mammals of Arabia*, Vol. II. Carnivora, Hyracoidea, Artiodactyla. Ernest Benn Ltd., London, xiv + 193–381.

Harrison, D.L., 1972. *The Mammals of Arabia*, Vol. III. Lagomorpha and Rodentia. Ernest Benn Ltd., London, xvii + 384–670 pp.

Harrison D.L. & P.J.J. Bates, 1991. *The Mammals of Arabia*. Harrison Zoological Museum, Kent, England, xvi +354 pp.

Hart, H. 1891. Some accounts of the fauna and flora of Sinai, Petra and Wadi Arabah. *Palestine Exploration Fund*, London, pp. 255.

Hasselquist, F., 1757. *Iter Palaestinum* in C. Linnaeus (ed.), Stockholm. p. 619.

Hatt, R., 1958. *The Mammals of Iraq*. Mus. of Zool., Univ. of Michigan. Misc. Public. No. 106, Ann Arbor, pp. 1–116.

Ivanitskaya, E., I Gorlov, O. Gorlova & E. Nevo, 1996. Chromosome markers for *Mus macedonicus* (Rodentia, Muridae) from Israel. *Hereditas* 124: 145–150.

Ivanitskaya, E., G. Shenbrot & E. Nevo, 1996. *Crocidura ramona* sp. nov. (Insectivora, Soricidae): a new species of shrew from the central Negev Desert, Israel. *Zeitschrift Saugetierkund* 61: 93–103.

Lay, D.M. 1967. A Study of the Mammals of Iran. *Fieldiana: Zool*. Oct. 31, Vol. 54. Field Mus. Nat. Hist. pp. 282

Lydekker, R., 1913. *Catalogue of the Ungulate Mammals in the British Museum*. (Nat. Hist.), Vol. 1. Brit. Mus. Pub. London.

MacCurdy, E., 1939. *The Notebooks of Leonardo de Vinci*. Reynal and Hitchcock, New York, pp. 1247.

Makin, D. & D.L. Harrison, 1988. Occurrence of *Pipistrelle ariel* Thomas, 1904 (Chiroptera: Vespertilionidae) in Israel. *Mammalia*, 52: 419–422.

Mayr. E. 1953. *Methods and Principles of Systematic Zoology*. McGraw Hill Book Co, Inc., New York.

Mayr, E. 1969. *Principles of Systematic Zoology*. McGraw Hill Book Co., New York. pp. 423.

Mendelssohn, H. 1974. The development of the population of Gazelles in Israel and their behavioral adaptations. In The behavior of ungulates and its relation to management. V. Geist and F. Walther (eds.) Morges: IUCN Pub. 722–739.

Mendelssohn, H. 1982. Wolves in Israel. pp. 173–195, in *Wolves of the World: Perspectives of Behavior, Ecology, and Conservation*. (F.H. Harrington and P.C. Paquet, eds.). Noyes Publication, Park Ridge, New Jersey, pp. 474.

Mendelssohn, H. & Y. Yom-Tov, 1987. *Mammals*. In A. Alon (ed.), *Plants and Animals of the Land of Israel. An Illustrated Encyclopedia*, [Hebrew] Ministry of Defense Publishing House and Society for the Protection of Nature, Vol. 7, Tel Aviv, pp. 295, Appendix pp. 111 in Hebrew.

Mendelssohn, H. & Y. Yom-Tov, 2000. *Mammalia of Israel*. The Israel Academy of Sciences and Humanities. Jerusalem, vii + 444 pp.

Miller, 1912. *Proc. U.S. Nat. Mus.*, 42: 171.

Nevo, E., 1961. Observations on Israeli populations of the mole rat, *Spalax l. ehrenbergi* Nehring, 1889. *Mammalia*, 25: 127–144.

Niethammer, J., 1959. Die nordafrikanischen Unterarten de Gartenschlafers (*Eliomys quercinus*). *Zeitschrift Saugetierkund*, 24: 35–45.

Ognev. S.I., 1928, 1931, 1935, 1940. *The Mammals of Eastern Europe and Northern Asia*. 4 Vol., Moscow.

Ognev. S.I., 1950. *Mammals of USSR and Adjacent Countries*. Vol. 7, Moscow.

Pocock, R.I., 1951. *Catalogue of the Genus* Felis. Trustees of the British Museum, London, pp. 190.

Qumsiyeh, M.B., 1996. *Mammals of the Holy Land*. Texas Tech. Univ. Press, Lubbock, pp. 389.

Ranck, G.L., 1968. The Rodents of Libya. Taxonomy, Ecology, and Zoological Relationships.

*United States Nat. Mus. Bull.*, 275. Smithsonian Inst. Pub. pp. 265.

Sclater, P.L. & O. Thomas, 1894–1900. *The Book of Antelopes*. Vols. 1–4. R.H. Porter Pub., London, pp. 242.

Shalmon, B., 1993. *A Field Guide to the Land Mammals of Israel: Their Tracks and Signs* [Hebrew]. Keter Pub. House Ltd., Jerusalem. 216.

Thorburn, A., 1920. *British Mammals*. Longmans, Green and Co. London. Two Vol. pp.

Tristram. G.H.B., 1884. *The Survey of Western Palestine. The Fauna and Flora of Palestine*. Comm. of the Pal. Explor. Fund. Pub. London, pp. 455.

Van Den Brink, F.H., 1955. *Die Saugetiere Europas*. Verlag Paul Parey, Hamburg-Berlin, pp. 225.

Wagner, A. 1839. Mammals collected by von Schubert on his journey to Egypt and Palestine in 1836–37. *Gelehrte Anzeigen* (Munich), 8: 297–300.

Wahrman, J. & A. Zahavi, 1955. Cytological contributions to the phylogeny and classification of the rodent genus *Gerbillus*. *Nature*, Vol. 175, p. 600,

Wahrman, J., R. Goitein & E. Nevo. Mole Rat *Spalax*: Evolutionary Significance of Chromosome Variation. *Science* 164: 82–84.

Walker, A. and P. Shipman, 1996. *The Wisdom of the Bones: In Search of Human Origins*, Alfred A Knopf, New York, p. 114.

Walker, E.P., 1964. *Mammals of the World*. The John Hopkins Press, Baltimore. Two Vol. pp. 1500.

Wassif, K. 1953. On a collection of mammals from Northern Sinai. *Bulletin Inst. Desert d'Egypt*, 3 (1): 107–118.

Wassif, K.& H. Hoogstall, 1954. The Mammals of South Sinai, Egypt. *Proceedings Egypt Acad. Sci.*, 9: 63–79.

Wiley, L.R., 1958. *Bible Animals: Mammals of the Bible*. Vantage Press, N.Y.

Yom-Tov, Y., 1967. On the taxonomic status of the hares (genus Lepus) in Israel. *Mammalia, 31*: 246–259.

Zafriri, A & S. Hellwing, 1973. The common shrew in Israel, *Crocidura russula monacha*: taxonomic aspects and data on reproduction under field conditions. *Isr. J. Zool.*, 22: 21.

Zahavi, A. & J. Wharman, 1957. The cytotaxonomy, ecology and evolution of the gerbils and jirds of Israel. (*Rodentia: Gerbillinae*). *Mamm. Paris.* 21: 341–380.

Zeuner, F.E., 1963. *A History of Domesticated Animals*. Hutchinson, London. pp. 560.

# INDEX OF NAMES

149

ferguson
©2000